Sonar for Practising Engineers

Third Edition

T0211476

Sonar for Practising Engineers

Third Edition

A. D. Waite

JOHN WILEY & SONS, LTD

Copyright © 2002   John Wiley & Sons Ltd,
The Atrium, Southern Gate, Chichester,
West Sussex PO19 8SQ, England

Telephone   (+44) 1243 779777

Email (for orders and customer service enquiries): cs-books@wiley.co.uk
Visit our Home Page on www.wileyeurope.com or www.wiley.com

Reprinted with corrections September 2005

This work is based on Sonar for Practising Engineers, Second Edition, published and distributed by Thales
Underwater Systems Limited (formerly named Thomson Marconi Sonar Limited), Ocean House,
Templecombe, Somerset, BA8 0DH (www.tms-sonar.com), 1998

Other Wiley Editorial Offices

John Wiley & Sons Inc., 111 River Street, Hoboken, NJ 07030, USA

Jossey-Bass, 989 Market Street, San Francisco, CA 94103-1741, USA

Wiley-VCH Verlag GmbH, Boschstr. 12, D-69469 Weinheim, Germany

John Wiley & Sons Australia Ltd, 33 Park Road, Milton, Queensland 4064, Australia

John Wiley & Sons (Asia) Pte Ltd, 2 Clementi Loop #02-01, Jin Xing Distripark, Singapore 129809

John Wiley & Sons Canada Ltd, 22 Worcester Road, Etobicoke, Ontario, Canada M9W 1L1

British Library Cataloguing in Publication Data
A catalogue record for this book is available from the British Library

ISBN 10: 0-471-49750-9 (P/B)
ISBN 13: 978-0-471-49750-9 (P/B)

To all my past and present colleagues within
the sonar community who have made
my career in sonar so rewarding
– and this book possible

Contents

Preface

Most books on sonar – the use of underwater sound for the detection, classification and location of underwater targets and for communications and telemetry – have been written by physicists and mathematicians. They are not always easily understood, nor are they immediately useful for solving the problems met by engineers and technicians.

The aims of this book – written by a practising engineer for practising engineers – are to provide an understanding of the basic principles of sonar and to develop formulae and rules of thumb for sonar design and performance analysis. No prior knowledge of sonar is assumed, and the physical principles and mathematics will be readily understood by engineers and technicians.

The earlier editions were produced to be supplied as back-up material to a short sonar course given by the author. This edition has been extensively rewritten to facilitate its use by an individual reader. Several new topics have been included:

- Echo sounding and side scan sonars for civil applications

- Communications sonars

- Low frequency active sonars

Many chapters contain worked examples and most chapters conclude with a few problems for the reader to solve; solutions are given at the end of the book. I hope these problems will be particularly useful to lecturers and students.

The book can be divided into three parts:

- *Equipment parameters*: this part briefly describes the motion of sound in an elastic medium, gives definitions of sound intensity and source level, explains the use of projector arrays to increase the source level, and looks at the use of hydrophone arrays to improve the signal-to-noise ratio of a wanted sound.

- *Propagation in the sea and backgrounds to detection*: this part considers spreading and absorption losses, propagation modes and simple modelling, noise and reverberation; it concludes with the sonar equations.

- *Practical sonar systems*: this part develops passive and active sonar systems; it looks at passive broadband, narrowband, intercept and communications sonars; and it considers active sonars for the detection of submarines, mines and torpedoes.

About the Author

Practising on HMS Jupiter!

Ashley Waite retired from the UK Ministry of Defence in 1990 after 40 years of Underwater Warfare Research at Portland.

Primarily engaged in Sonar Research and Development, his experience spans Surface Ship Sonars, Submarine Sonars, Surface Ship Torpedo Defence and a lesser involvement with Helicopter and Minehunting Sonars. He has also worked on Submarine Command Systems and the Underwater Aspects of Surface Ship Command Systems.

He then joined Ferranti Thomson Sonar Systems (now Thales Underwater Systems) as a Consultant on Active Sonar, working on improvements to the RN Sonar 2050; the active concepts of a fully integrated submarine sonar system (Sonar 2076); the Merlin helicopter sonar; and low frequency active sonar (Sonar 2087) for a further 10 years and still maintains an involvement with Sonar in a freelance capacity.

Introduction

Many methods of detecting the presence of underwater targets in the sea have been investigated. Here are some non-acoustic methods which have had varying degrees of success:

- Magnetic

- Optical signatures

- Electric field signatures

- Thermal detection (infrared)

- Hydrodynamic changes (pressure)

Magnetic methods include self-generated fields or perturbations of the earth's magnetic field, known as magnetic anomaly detection (MAD). Research continues into these methods but *underwater sound* is still unsurpassed, in spite of formidable difficulties facing its propagation through a highly variable medium and from the noise and reverberation backgrounds to detection. *Sonar* (sound navigation and ranging) – by analogy with *radar* (radio detection and ranging) – uses underwater sound for the detection, classification and location of underwater targets.

Passive sonar listens to the sound radiated by a target using a *hydrophone*, an underwater microphone, and detects signals against a background of the *ambient noise* of the sea and the *self-noise* of the sonar platform (an omnibus term to describe any vessel or site possessing a sonar system). Passive systems can be made directional, therefore the azimuth (horizontal bearing) of a signal is known. The nature of the signal – its frequency spectrum and how it varies with time – will help to classify the target.

Basic passive systems, however, give no information about the range of a target; a signal may belong to a close, quiet target or a noisy, distant target. More complex passive systems estimate range by the following methods:

- *Triangulation*: measuring the bearings of a target from two well-separated arrays.

- *Horizontal direct passive ranging (HDPR)*: based on the measurement of wavefront curvature using three well-separated arrays.

- *Vertical direct passive ranging (VDPR)*: measuring the vertical arrival angles of signals arriving at the same array via multiple paths as well as measuring the time differences between them.

All of these methods are fundamentally dependent on the accuracy of the bearing measurements and therefore demand large arrays and large separations to achieve useful range estimates.

Active sonar uses a *projector* (an underwater loudspeaker) to generate a pulse of sound which travels through the water to a target and is returned as an echo to a *hydrophone*, often the same device as the projector and in this context more commonly known as a *transducer*. The echo now has to be detected against a background of *noise* and *reverberation* (unwanted echoes from the sea surface and sea bed and from scatterers within the volume of the sea). Because the time between transmission of a pulse and reception of an echo can be measured and the speed of sound in the sea is known, the range of the echoing target is simply calculated. Active sonars are sometimes known as *echo ranging* systems.

To survive in sonar, newcomers must become familiar with the *decibel*. They will hear old hands discussing the design and performance of sonar systems in exchanges where almost every other word seems to be 'deebee' (dB or decibel). What is this decibel?

First of all, the bel is inconveniently large so it has been divided by 10 to become the decibel. It simply compares the power or intensity of the sound at one point in a system with that at another. The decibel defined:

$$\text{Power gain} = 10\log_{10}\left(\frac{P_{\text{out}}}{P_{\text{in}}}\right) \quad (\text{dB})$$

The decibel is of course also used in electronics, communications, radar and airborne sound, and will be familiar to most engineers. But it is in sonar where it really thrives. Long expressions such as the *sonar equations* are assembled with many parameters all expressed in decibel form. And to avoid serious errors, the decibel must be carefully defined and correctly applied to all of the terms in the equations.

Power ratios in sonar systems are frequently very large numbers. Calculations are greatly simplified when very large numbers are expressed in logarithmic form, so that values can be added instead of multiplied. Interestingly, although radar engineers may use decibels to describe the terms in a radar system (e.g. antenna gain, receiver noise factor), the radar equation is more commonly written in linear form rather than logarithmic form:

$$\text{SNR} = \frac{P_r}{N} = \frac{P_t G_t G_r \sigma \lambda^2 L_s}{(4\pi)^3 R^4 N}$$

The individual terms may be given their linear values and multiplied, or expressed in decibels and added.

Compare with an active sonar equation:

$$2\text{PL} = \text{SL} + \text{TS} - N + \text{DI} + 10 \log T - 5 \log d$$

All terms are in dB form; here are some possible values:

- PL, the propagation loss, might be 80 dB

- SL, the projector source level, might be 210 dB

- TS, target strength, might be 0 dB

All the examples, here and in the rest of the book, will use credible values for these parameters, and the reader should thereby learn to recognize and question suspiciously large or small values. (None of the terms in the sonar equations is likely to have a value outside the range −100 dB to +250 dB, at the very most.)

The *speed of sound* in water is about 1500 m/s (3355 mph), much faster than its speed in air of 340 m/s (760 mph) at sea level, but very much slower than the speed of light (electromagnetic radiation). Velocity, frequency and wavelength are related by

$$\boxed{c = f\lambda}$$

where

c = velocity in metres per second (m/s)

f = frequency in hertz (Hz) or cycles per second

λ = wavelength in metres (m)

Wavelength is the distance travelled by the wavefront during one cycle. At 1000 Hz we have $\lambda = 1.5$ m, and at 10 kHz we have $\lambda = 150$ mm. The speed of light (EM waves) is 3×10^8 m/s. At a typical radar frequency of 2000 MHz the wavelength is $\lambda = 150$ mm. Sizes of transmitting and receiving arrays, which typically have dimensions of several wavelengths, are therefore comparable for many sonar and radar systems.

Why Is Radar Not Used to Detect Underwater Targets?

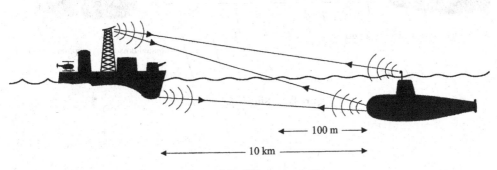

Figure 0.1 Why not radar?

The *radar* will detect the submarine's periscope and, assuming that the electro-magnetic energy spreads spherically, the inverse square law will apply and the EM propagation loss (two-way) is $40 \log 10\,000 = 160$ dB.

The *sonar* will detect the submarine's hull and, again assuming spherical spreading, the sound propagation loss is also 160 dB.

Therefore, given echoing surfaces above and below water, radar and sonar systems have comparable performances. But if the submarine is completely submerged, the EM losses over the final, say, 100 m are completely prohibitive at any radar frequency (Figure 0.1).

EM losses in the highly conductive sea water are given by $1400 f^{1/2}$ dB/km, where f is in kilohertz and, for just a 100 m path in the sea, the losses at 2000 MHz are 200 000 dB! Even at 30 kHz the losses equal 770 dB; the sea effectively presents a short circuit to the EM energy.

A modern nuclear-powered submarine can remain completely submerged for indefinite periods and during a mission may never expose any reflecting surface above water, but will rely completely upon its sophisticated sonars to navigate, to build up a complete acoustic scenario of its surroundings, to deploy counter-measures, and finally to prosecute an attack. Radar systems, therefore, will never be given opportunities to detect the submarine.

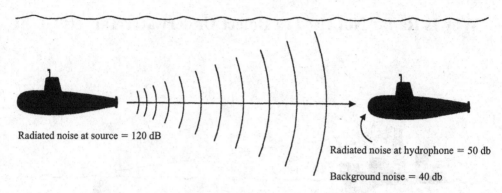

Figure 0.2   Simple passive sonar

The simplest passive sonar will have an omnidirectional wideband hydrophone and its detection performance is given by the *allowable propagation loss* (PL) between target and hydrophone (Figure 0.2). If we assume that, to make a detection, the radiated noise at the hydrophone must be 10 dB greater than the background noise, then

$$PL = 120 - 50 = 70 \text{ dB}$$

and assuming spherical spreading and no losses,

$$20 \log R = 70 \text{ dB}$$

therefore $R = 3000$ m.

Practical passive sonars will have directional arrays and limited operating bandwidths to reduce background noise, and they will integrate the signal (the radiated noise from a target) over a period of time. These measures will all improve the allowable PL significantly and greater ranges are possible. (But 120 dB represents quite a noisy submarine, so perhaps 3000 m is not too unrealistic a range.)

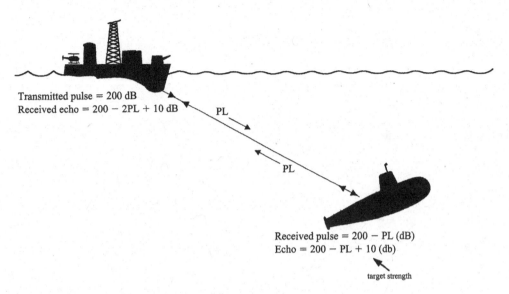

Transmitted pulse = 200 dB
Received echo = 200 − 2PL + 10 dB

PL

PL

Received pulse = 200 − PL (dB)
Echo = 200 − PL + 10 (db)

target strength

Figure 0.3 Simple active sonar

The simplest active sonar will have an omnidirectional projector and hydrophone (Figure 0.3). Its detection performance is given by the allowable two-way propagation loss (2PL), i.e. from projector to target and back to the hydrophone. If now the echo at the projector is 10 dB greater than the background noise, which for a surface ship sonar might be 60 dB, say, then

$$10 = (200 - 2\,PL + 10) - 60$$

$$PL = 70 \text{ dB}$$

therefore $R = 3000$ m.

Practical active sonars will have directional transmit and receive arrays, and a knowledge of the pulse will allow the processing to be matched to the returned echo. These measures will all improve the allowable PL significantly, and much greater ranges will be achievable.

1

Sound

1.1 Wave Motion

Sound is produced when an object vibrates and communicates its motion to the surrounding medium (Figure 1.1). Consider a regularly vibrating sphere; as it vibrates it alternately compresses and rarefies the surrounding medium, resulting in a series of compressions and rarefactions travelling away from the sphere. These are known as *longitudinal waves* since the particles move in the same dimension as the travelling wave. (Transverse wave motion occurs, for example, in a vibrating string, where the motion of the string is orthogonal to the direction of wave travel.)

1.2 Sound Pressure

For a wave to be a *plane* wave, pressure changes only in the direction of propagation of the sound; pressure is the same at all points in any plane normal to this direction. *Wavefronts* are those normal planes – separated by one wavelength, λ – where p is at a maximum.

The speed of sound refers to the longitudinal motion of the wavefronts in the medium and is related to wavelength and frequency by

$$c = f\lambda$$

The speed of sound is not to be confused with the *particle velocity*, u, which refers to the movement of the molecules in the medium.

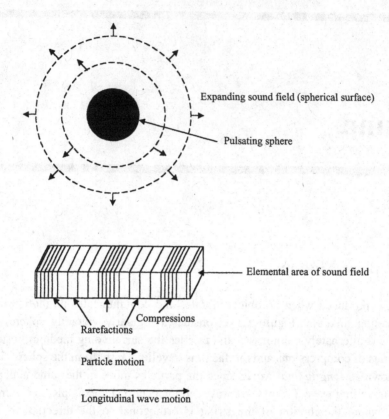

Figure 1.1 Propagation of sound in an elastic medium

Given a plane wave $\qquad\boxed{p = (\rho c)u}$

where

$p = pressure$ (Pa or N/m^2)

$u =$ particle velocity (m/s)

$\rho =$ fluid density $= 10^3$ kg/m^3 for sea water

$c =$ velocity of sound wave propagation

$\quad = 1.5 \times 10^3$ m/s in sea water

$\rho c =$ specific acoustic impedance, Z

$\quad = 1.5 \times 10^6$ kg m^{-2} s^{-1} for sea water

The sound wave carries mechanical energy with it in the form of the kinetic energy of the particles and the potential energy of the stresses in the medium. Because the wave is propagating, a certain amount of energy per second will flow across unit area normal to the direction of propagation.

This energy per second (power) crossing unit area is known as the *intensity* of the wave (power per unit area). For a plane wave, the intensity is related to the pressure by

$$I = p^2/\rho c$$

1.3   Reference Intensity

The reference intensity (I_r) in underwater sound is the intensity of a plane wave having a root mean square (RMS) pressure equal to 1 μPa (one micropascal). Inserting $p = 10^{-6}$ and $\rho c = 1.5 \times 10^6$ in the above equation for I, we obtain

$$I_r = 0.67 \times 10^{-18} \, \text{W/m}^2$$

Intensities are often loosely stated as 're 1 μPa'. This is clearly incorrect since the micropascal is a unit of pressure not intensity (power per unit area). Strictly, intensities should be stated as 're the intensity due to a pressure of 1 μPa'.

1.4   Source Level

The source level (SL) is defined as

$$SL = 10 \log \left(\frac{\text{intensity of source at standard range}}{\text{reference intensity}} \right)$$

The SL of an omnidirectional projector is always referred to a standard range (1 metre or 1 yard) from its acoustic centre. At 1 metre the acoustic centre of an omnidirectional source is surrounded by a sphere of surface area $4\pi r^2 = 12.6 \text{ m}^2$. If the omnidirectional power output is P watts, then the source intensity at 1 metre is $P/12.6 \text{ W/m}^2$ and SL becomes

$$SL = 10 \log \left(\frac{I_1}{I_r} \right)$$

$$= 10 \log \left(\frac{P/12.6}{0.67 \times 10^{-18}} \right)$$

$$= 10 \log P + 10 \log (1.1846 \times 10^{17})$$

$$= 10 \log P + 170.8 \text{ dB}$$

(If the standard range is 1 yard, then $SL = 10 \log P + 171.5 \text{ dB}$.)

If the projector is directional, then

$$DI_t = 10 \log \left(\frac{I_{dir}}{I_{omni}} \right)$$

where

DI_t = transmit directivity index

I_{dir} = intensity along the axis of the beam pattern

I_{omni} = intensity of the equivalent non-directional projector

and then SL becomes

$$SL = 10 \log P + 170.8 + DI_t$$

1.5   Radiated Power

P is the total *acoustic* power radiated by the projector, which is less than the *electrical* power supplied to it, P_e, and the ratio of these is the projector efficiency, E.

The efficiency depends on the bandwidth and may vary from as little as 0.2 to as high as 0.7 for a tuned, narrow bandwidth projector. The radiated powers for typical sonars may range from, say, 1 W to 40 kW and have DI_t values of between 10 and 20 dB.

Extremes of SL are therefore

$$SL = 10 \log 1 + 170.8 + 10 = 181 \text{ dB}$$

and

$$SL = 10 \log 40\,000 + 170.8 + 20 = 237 \text{ dB}$$

1.6   Limitations to Sonar Power

To achieve maximum range with active sonars, we need to generate the maximum amount of acoustic power – at least until the reverberation background limits the detection range. Against this purely technical argument, however, it should be remembered that when the power level is already high, say 20 kW, it may not be cost-effective to double the power to the technical limit since the additional 3 dB – which for active sonars means only 1.5 dB more in the allowable one-way propagation loss – may only increase the range by a small percentage. Attempts to increase the radiated power are eventually limited by two effects: cavitation and interaction.

1.7   Cavitation

When the power applied to a projector or an array is increased, bubbles form on the surface and there is a resultant loss in power by absorption and scattering within the bubbles, degradation of the beam pattern and a reduction in the acoustic impedance into which the projector generates (resulting in a mismatch with the transmitters supplying the projector array). Cavitation is a function of depth (pressure) and can be avoided by not exceeding the cavitation threshold (Figure 1.2).

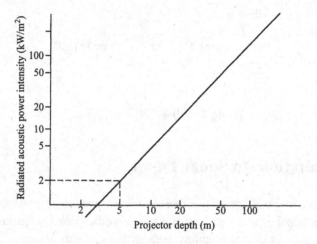

Figure 1.2   Cavitation threshold as a function of depth

Figure 1.2 gives a conservative estimate of the cavitation threshold for frequencies up to about 10 kHz and for pulses of at least 5 ms duration. It predicts the *start* of cavitation at some point before it spreads to the total radiating surface. In practice, higher power intensities are possible, perhaps 3 or 4 times higher, but these should be confirmed by experiment before finalizing a design.

Because cavitation takes a finite time to build up, albeit very short, the threshold also changes with pulse length and frequency:

- *Pulse length*: the threshold is increased for very short pulses, up to about 3 times for a 0.5 ms pulse compared with its 10 ms value, but is sensibly constant for longer durations.

- *Frequency*: the threshold is little changed up to about 10 kHz, but between 10 kHz and, say, 300 kHz the threshold increases roughly linearly with frequency.

Example 1.1

What is the maximum radiated power for a 100 ms pulse of frequency 10 kHz to avoid cavitation, for a cylindrical array of 1 m diameter and 1 m height, containing circular elements closely packed and operating at a depth of 5 m?

$$\text{Radiating surface area} = (\pi \times 1 \times 1) \times (\pi/4) = 2.5 \text{ m}^2$$

$$\text{Maximum radiated power} = 2.5 \times 2 = 5 \text{ kW}$$

And assuming $DI_t = 10$ dB, the source level is

$$SL = 10 \log 5000 + 171 + 10 = 218 \text{ dB}$$

Example 1.2

What is the maximum radiated power at 50 kHz to avoid cavitation at a depth of 5 m for a planar array of radiating surface 0.1 m^2 and pulse length 0.5 ms?

The radiated power intensity may be increased 3 times for the short pulse and 5 times for the higher frequency. Therefore

$$\text{Maximum radiated power} = 0.1 \times 2 \times 3 \times 5 = 3 \text{ kW}$$

And assuming $DI_t = 20$ dB, the source level is

$$SL = 10 \log 3000 + 171 + 20 = 226 \text{ dB}$$

Example 1.3

What is the maximum radiated power at 200 kHz to avoid cavitation at a depth of 50 m for a square array of radiating surface 0.01 m^2 and pulse length 0.5 ms? At this frequency, the array has a small size of 100 mm $\times$ 100 mm, which is $13\lambda \times 13\lambda$.

The radiated power intensity may be increased 3 times for the short pulse and 20 times for the higher frequency. Therefore

$$\text{Maximum radiated power} = 0.01 \times 50 \times 3 \times 20 = 30 \text{ kW}$$

And assuming $DI_t = 20$ dB, the source level is

$$SL = 10 \log 30\,000 + 171 + 20 = 236 \text{ dB}$$

Note, however, that the power intensity is 3000 kW/m^2 and for even moderate losses, say 20 per cent, the array must dissipate some 600 kW of heat per square metre or, to lapse into imperial, 6 kW over a 4 in square! Clearly the limiting factor at higher frequencies and/or greater depths is not cavitation but the power-handling capability of the array.

1.8   Interaction

When a number of projectors are assembled together in an array and driven electrically, the velocity of motion of the individual elements is not constant, but varies from element to element in a complex manner due to the acoustic interactions between them.

Unless this is compensated in the design, the interactive effects will reduce the total power output; affect the transmitted beam pattern; and produce a mismatch with the transmitters, which could damage transmitters, projectors or both. Interactive effects may be reduced in three ways:

- *Separating the elements of the array* produces a larger than optimum array and, particularly if the spacing is much greater than $\lambda/2$, deterioration in the transmit beam pattern. This may be acceptable if the resultant transmitter design problem is eased and total costs reduced.

- *Making the individual elements large* so that their self-radiation impedances are much greater than the mutual radiation impedances between elements.

- *Using individual amplifiers* to drive each element at the correct amplitude and phase to yield a uniform velocity of motion across the array.

1.9   Changes to Arrays

The problems of cavitation and interaction are far from trivial and have largely ensured the survival and renewed application of designs which, to the reader, must appear historical.

For example, the projector used in Sonar 2001 (an active and passive sonar system fitted to the Royal Navy's first nuclear-powered submarine, *HMS Dreadnought*) and still being considered for new sonars, was originally designed, tested and proven in the 1950s. The projector used in Royal Navy surface ship sonars 2016 and 2050 dates from 1972 and is highly likely to see service well into the twenty-first century.

Modular changes, however, such as halving the height or increasing the diameter of existing arrays, which preserve the element spacings and power intensity, or moderate frequency changes with appropriate scaling of dimensions, are likely to be successful without demanding long and risky testing.

1.10   Projector Sensitivity

If the voltage at the terminals of a projector is v, then the response S_V in db/V is given by

$$S_V = 10 \log \left(\frac{I_1}{I_r} \frac{1}{v^2} \right)$$

$$= SL - 20 \log v$$

In terms of power P, the response S_W in dB/W is given by

$$S_W = SL - 10 \log P$$

Example 1.4
If $SL = 200$ dB re 1 µPa (strictly, re the intensity of a plane wave of pressure 1 µPa) and $P = 30$ W, then $S_W = 200 - 15 = 185$ dB/W.

1.11   Hydrophone Sensitivity

If the sound pressure in micropascals at the hydrophone is p and the voltage at the open circuit terminals of the device is v, the hydrophone sensitivity is given by

$$S_h = 20 \log(v/p) = 20 \log v - 20 \log p \qquad (dB/V)$$

Example 1.5
If $20 \log p = 80$ dB re 1 Pa and $v = 1$ µV, then $S_h = -120 - 80 = -200$ dB/V. To find the output voltage from such a hydrophone, we use

$$20 \log v = S_h + 20 \log p$$

If $20 \log p = 100$ dB, then

$$20 \log v = -200 + 100 = -100 \text{ dB}$$

$$v = 10 \text{ µV}$$

1.12   Spectrum Level

The *spectrum level* (SpL) is the sound level in a 1 Hz band. It is expressed in dB relative to the intensity resulting from a sound pressure of 1 μPa. The *band level* (BL) refers to the total intensity of the sound in a band.

If the spectrum is flat (white noise) then

$$BL = SpL + 10 \log \Delta f$$

Example 1.6

If the spectrum level is 40 dB and $\Delta f = 3000$ Hz, then BL $= 40 + 35 = 75$ dB.

If the spectrum level is not flat, the band level can be obtained by integrating the intensity over the complete band. For ambient sea noise, the intensity may fall by 6 dB per octave. The total intensity is then given by

$$I_t = \int_{f_1}^{f_2} \frac{I_0}{f^2} df = I_0 \left[-\frac{1}{f} \right]_{f_1}^{f_2}$$

Example 1.7

What is the sound level (total intensity), BL, of sea noise in a band from 100 Hz to 10 kHz, given that the spectrum level at 100 Hz is 100 dB?

$$I_0 = I_{100} \times 10^4 \qquad I_t = I_{100} \times 10^4 \left[-\frac{1}{10^4} + \frac{1}{10^2} \right]$$

$$= I_{100} \times 10^2$$

and in dB form, BL $= 100 + 20 = 120$ dB. This is 20 dB less than if the spectrum were flat.

It is seldom necessary to calculate BL for a non-flat spectrum. A practical sonar receiver will either equalize (pre-whiten) its input, or process (simultaneously) bands narrow enough for the spectra to be assumed flat.

1.13   Sound in Air and in Sea Water

Note that in Tables 1.1 and 1.2 the velocity of sound – the longitudinal wave motion – is almost constant for a given medium, whereas the particle velocity is directly proportional to the pressure.

Table 1.1   Sound parameters in air

SOUND IN AIR $\rho c = 415$ kg m^{-2} s^{-1}

Sound level	Intensity I (W/m^2)	Pressure p (N/m^2)	Particle velocity u (m/s)	Particle displacement u/ω (m at 800 Hz)
Threshold of hearing 0 dB re 20 µPa	10^{-12}	2×10^{-5}	5×10^{-8}	10^{-11}
Conversation 60 dB re 20 µPa	10^{-6}	2×10^{-2}	5×10^{-5}	10^{-8}
Threshold of pain 120 dB re 20 µPa	1	20	5×10^{-2}	10^{-5}

Table 1.2   Sound parameters in sea water

UNDERWATER SOUND $\rho c = 1.5 \times 10^6$ kg m^{-2} s^{-1}

Sound level	Intensity I (W/m^2)	Pressure p (N/m^2)	Particle velocity u (m/s)	Particle displacement u/ω (m at 10 Hz)
DSS2 at 10 kHz 40 dB re 1 µPa	0.67×10^{-14}	10^{-4}	6.7×10^{-11}	1.1×10^{-15}
Typical reverb at 2 km range 120 dB re 1 µPa	0.67×10^{-6}	1	6.7×10^{-7}	1.1×10^{-11}
Array: source level 220 dB re 1 µPa	0.67×10^{-4}	10^{-5} (1 atm)	6.7×10^{-2}	1.1×10^{-6}

Note: it is interesting that 1.1×10^{-15} is only 3 millionths of the diameter of a hydrogen molecule (0.28 nm)

1.14 Problems

1.1 The pressure, p, of an underwater sound is 100 µPa. What is the intensity, I, of the sound and its level expressed in decibels compared with the reference intensity, I_r?

1.2 What is the source level, SL, of a projector radiating 40 kW of acoustic power and with a directivity index, DI_t, of 15 dB?

1.3 What is the safe maximum radiated power to avoid cavitation, given a 100 ms pulse of frequency 20 kHz and a planar array of 2 m length by 1 m height, containing circular elements closely packed and operating at a depth of 10 m?

1.4 The total intensity of the sound in an octave from 2000 to 4000 Hz is 80 dB relative to the intensity resulting from a sound pressure of 1 µPa. Assuming the band is flat, what is the spectrum level, SpL, of the sound?

2

Arrays

2.1   Need for Projector Arrays

Transducers operating in the transmit mode – *projectors* – are assembled as arrays in order to increase the *source level* of the transmitted acoustic pulse. The *directivity* of a projector array concentrates the transmitted sound in a given direction.

In the horizontal plane the transmissions may be omnidirectional or, with a corresponding increase in source level, they may take place over some smaller sector. In the vertical plane there is clearly no point in transmitting omnidirectionally, so a typical array is designed to make the vertical beamwidth between 3° and 30° to the 3 dB points.

The required source level is seldom available from one projector, except perhaps at very low frequencies where the surface area is likely to be quite large. A single projector requires a large surface area to avoid cavitation, and at higher frequencies this is not a simple engineering solution: directivity would be sacrificed or at best compromised. Finally, the reliability of an array of projectors is much better because several, perhaps even half, of the projectors can fail before sonar performance is critically impaired.

2.2   Need for Hydrophone Arrays

Transducers operating in the receive mode – *hydrophones* – are assembled as arrays to improve the response of the array in a desired direction, thereby increasing the signal-to-noise ratio and indicating the direction of a signal source. The *directivity* of a hydrophone array is due to all signals impinging upon the array from the same direction being in phase, and therefore reinforcing one another.

2.3   Beam Patterns

The *beam pattern* of an array plots the array response against angle, describing the relative responses of a beam formed from an array to signals and noise from all directions. Mathematically, the beam pattern of an array is identical for both transmit and receive. The improvement in the signal-to-noise ratio due to the array is known as the *array gain*. A special case of array gain, where the signal is coherent and the noise is incoherent, is the *directivity index*. This parameter is simpler to calculate and will normally be a satisfactory measure of the increase in the signal-to-noise ratio due to the array.

2.4 Directivity of a Dipole

S_1 and S_2 are two point sources (forming a dipole) vibrating in phase and at equal amplitudes (Figure 2.1). A hydrophone placed in the far field in the direction θ receives two sound waves of equal amplitude (since S_1 and S_2 are almost equidistant from the hydrophone), but of different phase corresponding to the distance $S_2 H$:

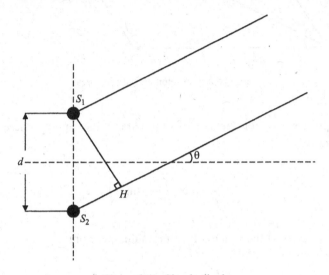

Figure 2.1   Simple dipole

$$S_2 H = d \sin \theta$$

and the phase difference is

$$\varphi = \frac{2\pi d}{\lambda} \sin \theta$$

The sound pressure at the hydrophone is the vector sum of the pressures p_1 and p_2, of the same amplitude and with a phase difference φ (Figure 2.2):

$$p_1 + p_2 = 2 p_1 \cos \frac{\varphi}{2}$$

For a single source S_1, the pressure is constant whatever the direction. When a second source S_2 is present, the pressure varies between 0 and $2p$ as a function of θ:

$$p_\theta = 2 p_0 \cos \frac{\varphi}{2} = 2 p_0 \cos \left(\frac{\pi d}{\lambda} \sin \theta \right)$$

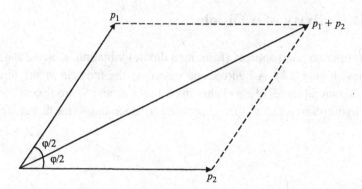

Figure 2.2   Vector sum of pressures

$2p_0$ is the pressure, p_a, on the axis of symmetry of the dipole (where $\theta = 0$) and we can write

$$\frac{p_\theta}{p_a} = \cos\left(\frac{\pi d}{\lambda}\sin\theta\right)$$

The intensity varies as the square of the pressure, therefore

$$\frac{I_\theta}{I_a} = \cos^2\left(\frac{\pi d}{\lambda}\sin\theta\right)$$

and in dB form, we have

$$\boxed{10\log\left(\frac{I_\theta}{I_a}\right) = 20\log\left[\cos\left(\frac{\pi d}{\lambda}\sin\theta\right)\right]}$$

The 3 dB beamwidth (the width in degrees at the half-power points of the dipole response) for different spacing of the dipole elements is obtained by putting

$$20\log\left[\cos\left(\frac{\pi d}{\lambda}\sin\theta\right)\right] = 3\text{ dB}$$

$$\cos\left(\frac{\pi d}{\lambda}\sin\theta\right) = \frac{\sqrt{2}}{2}$$

$$\frac{\pi d}{\lambda}\sin\theta = \frac{\pi}{4}$$

Table 2.1 gives the 3 dB beamwidth at four element spacings and Figure 2.3 shows the beam patterns. When the spacing is exactly $\lambda/2$, nulls appear at 90° and 270°. As the spacing is increased, secondaries (sidelobes) appear which are soon unacceptable, and when $d = \lambda$ they totally change the beam pattern so that the major responses now occur at the previous null angles.

Table 2.1 3 dB beamwidths

Element spacing, d (wavelength units)	θ (degree)	3 dB beamwidth (degree)
$\lambda/4$	90	180
$\lambda/2$	30	60
$3\lambda/4$	19.5	39
λ	14.5	

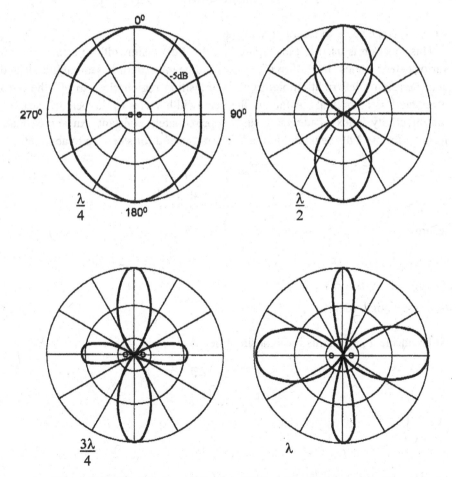

Figure 2.3 Dipole beam patterns as a function of spacing in wavelengths

2.5   The General Line Array

When the simple dipole theory is extended to an array of n hydrophones of spacing d, the beam pattern is given by

$$b(\theta) = 20 \log \left[\frac{\sin\{(n\pi d/\lambda)\sin\theta\}}{n \sin\{(\pi d/\lambda)\sin\theta\}} \right]$$

The main lobes of an array can be steered by introducing phase or time delays in series with the elements. When this is done the beamwidth and sidelobe structure of the array is changed. The beam pattern of the line array is then modified to

$$b(\theta) = 20 \log \left[\frac{\sin n\pi (d\sin\theta/\lambda - d\sin\theta_s/\lambda)}{n \sin \pi (d\sin\theta/\lambda - d\sin\theta_s/\lambda)} \right]$$

where θ_s is the angle of steer at broadside.

This function is used to plot Figure 2.4, the 3 dB beamwidth of a line array at various steer angles. In endfire ($\theta_s = 90°$) the beam pattern is searchlight shaped and there is no left/right ambiguity. In a transition region close to 90° the 3 dB beamwidth is ambiguous. Although the beamwidth varies as the beam is steered, the directivity index of the array stays approximately constant. An approximate formula for the 3 dB beamwidth of a line array, useful for steers up to about 60°, is

$$2\theta_3 = \frac{76}{Lf} \left(1 + \frac{\theta_s^2}{4000} \right)$$

where

$L = $ length (m)

$f = $ frequency (kHz)

$\theta_s = $ steer angle (degree)

At broadside, $\theta_s = 0°$, this formula simplifies to

$$2\theta_3 = \frac{76}{Lf}$$

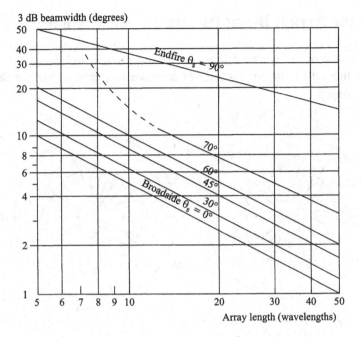

Figure 2.4 Beamwidth of a line array of $\lambda/2$ spaced elements at various steer angles

2.6   Line Array: Beam Pattern vs. Steer Angle

The beam patterns in Figure 2.5 illustrate the transition from broadside to endfire. Note that the 3 dB ambiguity is already apparent at a steer of 60° for this short (5λ) line array.

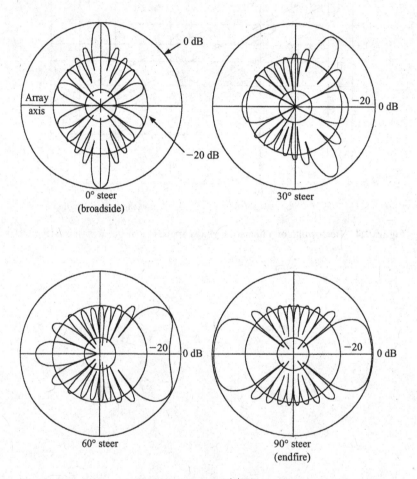

Figure 2.5   Line array, 10 elements spaced $\lambda/2$: beam patterns at various steers

2.7   Broadside Array: Length and Spacing

The beam patterns in Figure 2.6 further illustrate the effects of array length and element spacing at broadside. Note again the total change to the beam pattern as the spacing approaches λ.

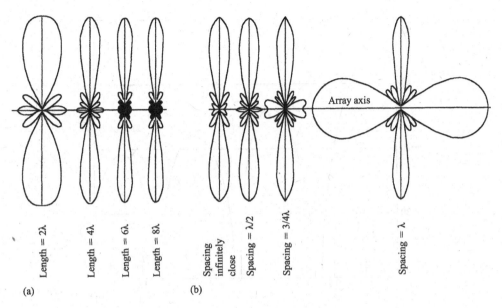

Figure 2.6 Effects of length and element spacing on a broadside array: (a) element spacing constant at $\lambda/2$, array length increasing; (b) array length constant at 3λ, element spacing increasing

2.8 Beam Pattern for a Continuous Line

For a continuous line of length L, the pressure in the far field is found by integration to be

$$\frac{p_\theta}{p_\alpha} = \frac{\sin\{(\pi L/\lambda)\sin\theta\}}{(\pi L/\lambda)\sin\theta}$$

This expression (Figure 2.7) is a sin x upon x function, where $x = (\pi L/\lambda)\sin\theta$; it is somewhat easier to handle than the earlier expression derived from the summation of n hydrophones of spacing d, an array whose length is given by $(n-1)d$.

The intensity of the beam pattern in dB form is

$$b(\theta) = 20\log\left[\frac{\sin\{(\pi L/\lambda)\sin\theta\}}{(\pi L/\lambda)\sin\theta}\right]$$

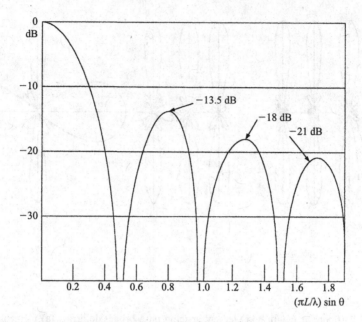

Figure 2.7   Normalized beam pattern for a continuous line

There are nulls in the beam pattern where $x = \pm\pi$, $\pm2\pi$, $\pm3\pi$, ..., and peaks in the beam pattern where $x = 0$, $\pm\frac{3}{2}\pi$, $\pm\frac{5}{2}\pi$, ...

The sidelobe levels are given by

$$20\log\left[\frac{2}{\pi(2m+1)}\right]$$

where m is the sidelobe number. The results apply quite well to an array of *spaced elements* provided their spacing, d, is not more than about $\lambda/2$ and the array is at least 3λ in length.

Table 2.2

x	$x^{-1}\sin x$	$20\log(x^{-1}\sin x)$
0	1	0
1.39	0.71	-3
2.27	0.32	-10
2.85	0.1	-20

The numerical values in Table 2.2 can be used to derive some useful approximations for the shape of the main lobe. Beginning with the 3 dB beamwidth (Hz):

$$1.39 = \frac{\pi L}{\lambda} \sin \theta_3$$

Converting to radians, for small angles we obtain

$$1.39 = \frac{\pi L}{\lambda} \theta_3 \left(\frac{\pi}{180} \right)$$

If the elements are spaced by $\lambda/2$, then

$$2\theta_3 = 100/n$$

Similar expressions can be produced for 10 dB and 20 dB beamwidths and the results are given in Table 2.3. Note that the 20 dB beamwidth is about double the 3 dB beamwidth.

Table 2.3 Beamwidth expressions

$2\theta_3$	$2\theta_{10}$	$2\theta_{20}$
$50\lambda/L$	$82\lambda/L$	$104\lambda/L$
$100/n$	$164/n$	$208/n$

Note: the formula $100/n$ is an alternative to the approximate formula of Section 2.5

Example 2.1
What are the 3 dB, 10 dB and 20 dB beamwidths for a line array of length 6λ? What are the bearings of the first-, second- and third-order sidelobes relative to the direction of maximum intensity?

$$2\theta_3 = 50\lambda/L = 8.3°$$

$$2\theta_{10} = 82\lambda/L = 13.7°$$

$$2\theta_{20} = 104\lambda/L = 17.3°$$

- First sidelobes occur where $\pm\frac{3}{2}\pi = 6\pi \sin \theta$, i.e., where $\theta = \pm14.5°$
- Second sidelobes occur where $\pm\frac{5}{2}\pi = 6\pi \sin \theta$, i.e., where $\theta = \pm25°$
- Third sidelobes occur where $\pm\frac{7}{2}\pi = 6\pi \sin \theta$, i.e., where $\theta = \pm36°$

2.9   Shading

Except for the simple dipole with element spacings less than or equal to $\lambda/2$, all
the above beam patterns have significant sidelobe levels. The first-, second- and
third-order sidelobes are respectively only 13.5, 18 and 21 dB below the peak level
of the main lobe. Strong signals will be detected through the sidelobes of adjacent
beams as well as, correctly, in the main lobe of the beam at the bearing of the
signal. The resultant bearing ambiguity and additional, false, signals complicate
all further processes. The aim, therefore, should be to produce the narrowest
possible main lobe consistent with some reasonable level of sidelobes.

By using amplitude shading or weighting, i.e., varying the amplitudes of the
signals applied across the array (in transmit) or coming from the array (in receive),
it is possible to reduce the sidelobes, but always at the expense of widening the
main lobe and reducing the DI somewhat. In transmit, reducing the DI means that
the source level (SL) is reduced as a consequence. In an active system it is the
combined, transmit and receive, beam pattern which is important and it is hardly
ever desirable to further reduce the sidelobes at the expense of SL – far better to
adequately shade the receiver elements only.

All shading functions widen the main lobe by some factor; the factor is 1.36 for
the cosine shading function in Figure 2.8. Note also that the skirts of the main lobe

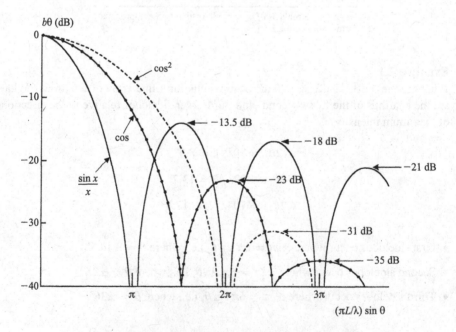

Figure 2.8   Normalized beam patterns: line array, elements spaced $\lambda/2$

extend into the first sidelobes of the *unshaded* pattern, and therefore simply quoting the level of the first sidelobe is an incomplete description and may be misleading (Section 12.6).

Mathematical procedures, based on work by Dolph and Chebyshev, determine the amplitude shading coefficients which yield the *narrowest main lobe for a specified level of sidelobes* (Figure 2.9). For example, the coefficients for a six-element line array to yield −30 dB sidelobes are 0.30, 0.69, 1, 1, 0.69, 0.30. And the main lobe is broadened, at the 3 dB points by a factor of 1.3.

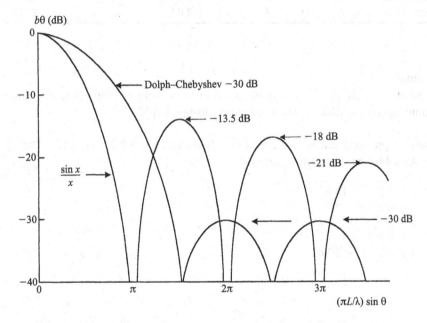

Figure 2.9   Unshaded and Dolph–Chebyshev shaded, normalized beam patterns

Some characteristics of commonly used weighting functions, normalized to the unweighted array, are given in Table 2.4. All dB values in the table are rounded to the nearest 0.5 dB. Note that all of the other weighting functions broaden the main lobe more than the Dolph–Chebyshev −30 dB weighting function. The theoretical sidelobe levels are generally lower but they are seldom fully achieved in practice. Element failures and, in the case of a towed line, curvature of the array, ensure that it is unsafe to expect sidelobes much better than, say, 25 dB down on the main lobe.

Note the similarity between amplitude shading used to reduce bearing sidelobes in a beam pattern and pulse shaping used to reduce frequency sidelobes in the frequency spectrum of a pulse, as described in Chapter 9.

Table 2.4 Commonly used weighting functions

Weighting function	Reduction in transmit power (dB)	Broadening factor, 3 dB	$10 \log BF_3$ (dB)	Broadening factor, 20 dB
None	0	1	0	1
Dolph–Chebyshev 30 dB	3	1.3	1	1.5
Cosine	3	1.4	1.5	1.6
Gaussian	3	1.4	1.5	1.6
Hamming	3.5	1.5	2	1.7
Squared cosine	3.5	1.6	2	1.9
Binomial	4	2.0	3	2.2

Example 2.2

If the array of the previous example is (a) Dolph–Chebyshev 30 dB shaded and (b) Hamming shaded, what are the 3 dB and 20 dB beamwidths?

From the previous example, the unshaded beamwidths are 8.3° and 17.3°. Therefore the shaded beamwidths are as follows:

(a) $\quad 1.3 \times 8.3° = 10.8°$
$\quad\quad 1.5 \times 17.3° = 26°$

(b) $\quad 1.5 \times 8.3° = 12.5°$
$\quad\quad 1.7 \times 17.3° = 29.4°$

2.10   Shaded Arrays: Transmit Source Levels

Shading a transmit array reduces the total output power (assuming that the power outputs of unity-weighted elements cannot be increased). The total power is also spread over a greater beamwidth and therefore the SL is significantly reduced compared to an unshaded array. A reduction in SL will reduce the noise-limited performance of an active sonar and should be avoided. However, if it is essential – perhaps to broaden the vertical beamwidth of a sonar to achieve adequate depth cover at close ranges – the reduced SL may be estimated quite simply using Table 2.4:

$$\text{SL reduction} = \text{total power reduction} + 10 \log BF_3$$

where BF_3 is the broadening factor at 3 dB.

Example 2.3

A Dolph–Chebyshev array designed to achieve −30 dB sidelobes will have its SL reduced due to the reduction in transmit power, 3 dB, and due to the broadening factor, 1 dB, i.e., a total reduction of 4 dB.

Check

Given a line array of six elements, spaced $\lambda/2$ and each capable of a maximum output of 1000 W omnidirectionally, we obtain

$$SL = 10 \log 6000 + 171 + 10 \log 6 = 217 \text{ dB}$$

If the line array is now amplitude weighted,

$$0.3, 0.69, 1, 1, 0.69, 0.3$$

the total power output is

$$90 + 480 + 1000 + 1000 + 480 + 90 = 3140 \text{ W}$$

and the power reduction is $10 \log(6000/3140) = 2.8$ dB.

Also, because the 3 dB beamwidth is increased by a factor of 1.3, this power is spread over a larger sector and the average power is reduced by $10 \log 1.3 = 1.1$ dB. The total reduction is therefore $2.8 + 1.1 = 3.9$ dB, which shows close agreement. The source level is now $217 - 4 = 213$ dB.

2.11 Directivity Index

When the hydrophones are assembled in arrays there is an improvement in the signal-to-noise ratio (S/N). This improvement is known as the *array gain* and is defined as

$$AG = 10 \log \frac{(S/N)_{\text{array}}}{(S/N)_{\text{element}}}$$

It is difficult to compute this quantity because it relies on knowing the coherences of signal and noise across the dimensions of the array.

An important case is where the signal is a plane wave and coherent, and the noise is isotropic and incoherent. Here the AG reduces to a more tractable and easily visualized quantity known as the *directivity index* (DI):

$$DI = 10 \log \frac{\text{peak intensity of radiated pattern}}{\text{average intensity of radiated pattern}}$$

We will first derive the directivity index for a simple dipole and, avoiding complex mathematical treatment, extend this result to cover the directivity indices for more complex arrays.

2.12 DI of a Simple Dipole

The dipole, elements S_1 and S_2, is at the centre of a large sphere of radius R (Figure 2.10). The sound flux passing through the shaded zone on the diagram is then

$$I_\theta(2\pi R \cos \theta)R \, d\theta$$

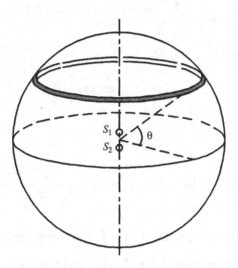

Figure 2.10   Directivity index of a simple dipole

If $\bar{I}$ is the average intensity at the surface of the sphere, the total sound flux passing through its complete surface is

$$\bar{I} \times 4\pi r^2$$

and therefore

$$\bar{I} = \frac{1}{4\pi r^2} \int_{-\pi/2}^{\pi/2} I_\theta(2\pi R^2 \cos \theta)d\theta$$

and the *directivity factor*, K (DI $= 10 \log K$), is given by

$$\frac{1}{K} = \frac{\bar{I}}{I_\alpha} = \frac{1}{4\pi R^2} \int_{-\pi/2}^{\pi/2} \frac{I_\theta}{I_\alpha}(2\pi R^2 \cos \theta)d\theta$$

and substituting

$$\frac{I_\theta}{I_\alpha} = \cos^2\left(\frac{\pi d}{\lambda}\sin\theta\right)$$

(see Section 2.4), we obtain

$$\frac{1}{K} = \frac{1}{2}\int_{-\pi/2}^{\pi/2}\cos^2\left(\frac{\pi d}{\lambda}\sin\theta\right)\cos\theta\,d\theta$$

Integrating by changing the variable, we obtain

$$\frac{1}{K} = \frac{1}{2}\left[1 + \frac{\sin(2\pi d/\lambda)}{2\pi d/\lambda}\right]$$

and

$$\text{DI} = 10\log K = 3 - 10\log\left[1 + \frac{\sin(2\pi d/\lambda)}{2\pi d/\lambda}\right]$$

- When the elements are separated by $d = \lambda/2$, this reduces to DI $= 3$ dB
- When the elements are separated by $d = 0$, then DI $= 0$ dB

We would expect both of these results intuitively: two elements are twice (3 dB) as good as one; and when the two become one ($d = 0$), the 'dipole' is omnidirectional and DI $= 0$ dB.

2.13 DI of a Line Array

For a line array of n elements spaced $\lambda/2$,

$$\boxed{DI = 10\log n}$$

Clearly the elements in the line are only spaced $\lambda/2$ at one frequency. Therefore, for a wideband system the DI will change with frequency. As the frequency is reduced, the line approximates to a continuous line of length $L = n\lambda_0/2$, where λ_0 is the wavelength at f_0, the design frequency for the line array and

$$\boxed{DI = 10\log\left(\frac{2L}{\lambda}\right)}$$

Note that the DI falls at a rate of 3 dB per octave.

An alternative expression for the DI of a line which will suffice for two or three octaves *below* the design frequency f_0 (i.e., at a frequency where $d = \lambda/2$) is

$$\boxed{DI = 10\log n + 10\log(f/f_0) \quad \text{(dB)}}$$

Note that, from its definition, DI is always positive and cannot fall below 0 dB.

As the frequency is increased, the line can no longer be considered as approximating to a continuous line, and excessive sidelobes appear as the spacing approaches λ. Therefore this equation should not be used for frequencies above, say, $1.5f_0$.

Example 2.4
A line consists of 64 elements spaced $\lambda/2$ at 1 kHz. What is its DI at 1 kHz, 100 Hz and 1.5 kHz?

Length of line $L = 32\lambda = 32 \times 1.5 = 48$ m

At 1 kHz $DI = 10\log n = 10\log 64 = 18$ dB

At 100 Hz $DI = 10\log(2L/\lambda) = 10\log(2 \times 48/15) = 8$ dB

At 1.5 kHz $DI = 18 + 10\log 1.5 = 19$ dB

2.14 DI of a Planar Array

The DI of a planar array is simply calculated by considering it as the sum of a number of linear arrays:

$$\boxed{\mathrm{DI} = 10 \log mn}$$

where m is the number of lines of n elements, all spaced by $\lambda/2$.

The array is usually baffled, as when mounted on the flank of a submarine. For a baffled array, add 3 dB to this result, so the wideband equation of the previous section then becomes

$$\boxed{\mathrm{DI} = 3 + 10 \log mn + 20 \log(f/f_0) \qquad \text{(dB)}}$$

Note again that DI is always positive and cannot fall below 0 dB. As the frequency is increased, the planar array can no longer be considered as approximating to a continuous surface, and excessive sidelobes appear as the spacing approaches λ. Therefore this equation, in common with the equation for a line array, should not be used for frequencies above, say, $1.5 f_0$.

Example 2.5
A flank array has 8 rows of 32 elements spaced $\lambda/2$ at 1 kHz. What are its DI at 800 Hz and its 3 dB beamwidths at 1 kHz and at steer angles of (a) broadside, (b) 30°, (c) 60°?

$$\boxed{\mathrm{DI} = 3 + 10 \log(8 \times 32) + 20 \log 0.8 = 25 \text{ dB}}$$

The length of the array is 16λ and its height is 4λ. The horizontal and vertical beamwidths are given by the same formula:

$$\boxed{\theta_3 = \frac{76}{Lf}\left(1 + \frac{\theta_s^2}{4000}\right)}$$

(a) Horizontal beamwidth $= 76/(24 \times 1) = 3.2°$
    Vertical beamwidth $\quad= 76/(6 \times 1) = 12.7°$

(b) $\theta_h = (76/24)(1 + 900/4000) = 3.9°$
    $\theta_v = 12.7°$ (steer is only horizontal)

(c) $\theta_h = (76/24)(1 + 3600/4000) = 6.0°$
    $\theta_v = 12.7°$ (steer is only horizontal)

2.15   DI of a Cylindrical Array

A cylindrical array of elements comprising a number of vertical lines (staves) is a good practical shape (Figure 2.11). The number of staves is simply related to the beam spacing and is also, conveniently, a binary number. This makes the beamforming simple because all beams are identical and provide 360° cover in azimuth. The horizontal beamwidth is constant and the vertical beamwidth is a function of the stave height.

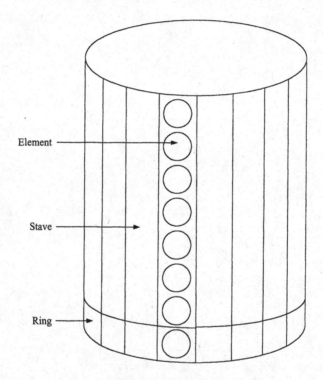

Figure 2.11   Cylindrical array

To form a beam, a certain number of staves – usually about one-third the periphery of the cylinder – are phased to make them appear as a planar array. Adjacent beams are then formed by stepping around the array at an indexing angle equal to the required beam spacing.

A practical formula for the DI of a baffled cylinder is

$$DI = 10 \log 5 h d f_0^2$$

where

h = height (m)

d = diameter (m)

f_0 = design frequency (kHz)

This is for a 'full' beam, i.e., a beam formed by using all the usefully contributing elements on the surface of the cylinder (an aperture of about 120° of the cylinder). The design frequency, f_0, is the frequency where the elements are spaced $\lambda/2$.

Example 2.6
A cylindrical array has height 1 m, diameter 2 m and design frequency 5 kHz, so $DI = 10\log(5 \times 1 \times 2 \times 25) = 24$ dB.

If 'half-beams' are formed for cross-correlation or phase comparisons (see later), the DI of each half-beam is 3 dB less than the DI of the corresponding full beam. Here are some practical formulae for the *beamwidths* of a cylinder.

Horizontal beamwidth

$$\theta_h = \frac{88}{df_0}$$

This formula uses one-third of the array's periphery, and we obtain

$$\theta_h = \frac{88}{2 \times 5} = 8.8°$$

Vertical beamwidth

$$\theta_v = \frac{76}{hf_0}$$

This is the same as the formula for line arrays, and we obtain

$$\theta_v = \frac{76}{1 \times 5} = 15.2°$$

2.16   DI Formulae for Simple Arrays

The above formulae for simple arrays are mostly in terms of numbers of elements, spaced $\lambda/2$. Because $n = 2L/\lambda$, these formulae may also be expressed in terms of the dimensions of the arrays. Table 2.5 lists the formulae for linear, planar, square, circular and cylindrical arrays in both forms. Some of these formulae are approximate but are perfectly adequate for practical sonar design.

Table 2.5   Formulae for DI

Array	Directivity index (dB)	
Line	$10 \log n$	$10 \log(2L/\lambda)$
Planar	$10 \log mn$	$10 \log(4Lh/\lambda^2)$
Square	$20 \log n$	$20 \log(2L/\lambda)$
Circular	$20 \log n$	$20 \log(1.77d/\lambda)$
Cylindrical, baffled (see notes)	$3 + 10 \log mn$	$10 \log(5hdf_0^2)$ or $10 \log(11hd/\lambda^2)$

Notes: m is the number of elements in one-third of a ring, n is the number of elements in a stave; f_0 is in kHz

2.17   Conformal Arrays

Conformal arrays follow the hull form of a vessel, usually a submarine. They have been fitted to single-hulled submarines, and in order to provide a good surveillance capability, they occupy both port and starboard flanks and extend around the bow. Estimates of DI and beamwidths (which change significantly with azimuth) can be made by approximating the array to an assembly of planar arrays (on the flanks where the curvature is not great) and part cylinders (on the bow).

2.18   Spherical Arrays

Spherical arrays are used where greater vertical coverage is required (or to compensate for ship movements). However, given the wasteful use of projectors or hydrophones to provide coverage in this manner, and the *reduced DI* compared with a cylindrical array of the same diameter, it is debatable whether adequate vertical cover is not better achieved by vertical steering of a cylindrical array – using complete staves rather than selecting elements as with a spherical array – and accepting the small losses in DI which then exist.

2.19 Volumetric Arrays

When the elements occupy three dimensions, e.g., in two or more concentric cylinders, the array is known as *volumetric* (Figure 2.12). It is then possible, by suitable separation and phasing between the cylinders, to generate directivity patterns and directivity indices equivalent to a baffled cylindrical array.

Consider an array of two concentric cylinders each made up of 16×6 element staves (Figure 2.12). The elements are separated by $\lambda/2$ and the cylinders are separated by $\lambda/4$. The design frequency is 5 kHz. The overall size of the array is therefore a cylinder of diameter 0.75 m and height 0.75 m. Staves A and B are separated by $\lambda/4$, and if the relative phase is 90°, the horizontal directivity of the pair of staves is then cardioid in shape, as shown. The resultant directivity index of the complete array is therefore increased by 3 dB and is the equivalent of a baffled cylindrical array given by

$$\begin{aligned}
\mathrm{DI} &= 10 \log 5hdf_0^2 \\
&= 10 \log(5 \times 0.75 \times 0.75 \times 25) \\
&= 18.5 \text{ dB}
\end{aligned}$$

Helicopter dipping sonars

Helicopter dipping sonars often use volumetric arrays. The array is mounted on ribs which fold (umbrella fashion) for stowage and lowering from the helicopter into the sea. When the array reaches the required depth, the rib fastenings are released and the array configures as a number of concentric cylinders. A separate transmitting array of omnidirectional projectors (omnidirectional in the horizontal plane) may complete the assembly.

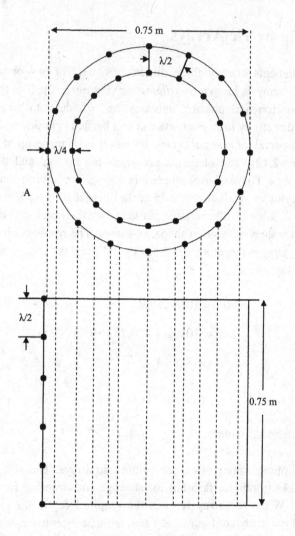

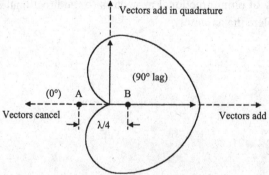

Figure 2.12   Volumetric, concentric cylinders, array

2.20   Beamformers

Analogue

Analogue beamformers are unlikely to be used except for very simple arrays in modern systems. They are assembled using lumped constant (LC) networks which must have a linear phase relationship with frequency. They are difficult to modify and cannot, at least not simply, be made to respond to sound speed changes. (When the speed of sound changes, because its frequency cannot change, its wavelength must change. Therefore the parameters of the beam will also change unless the speed of sound is used in the beamforming process.)

Digital

Digital beamformers use shift registers or random access memory store (RAM) to hold sampled data. The beams are then formed by addressing the samples. Shading, functions and sound speed variations are easily accommodated.

Narrowband systems

If the beamformer is for use with a narrowband system then phase delays, which are only exactly correct at one frequency, may be used to form the required beams instead of the more usual time delays which are constant with frequency and hence suitable for broadband systems.

2.21 Domes and Arrays

Arrays seldom have an ideal hydrodynamic shape and they are not sufficiently robust to survive without some protection. They are therefore normally housed in *domes* made of acoustically 'transparent' material. Modern domes are often built up using glass-fibre epoxy resin mixes, which give a good compromise between acoustic and mechanical properties. The acoustic 'windows' in the dome may simply be homogeneous, thinner areas or they may use a different material (e.g., rubber). No material transmits sound loss-free – some energy will be absorbed and some reflected. The design of *baffles*, which are used to attenuate unwanted noise from the propellers of a ship, for example, can also help to minimize the unwanted effects of reflections (Figure 2.13).

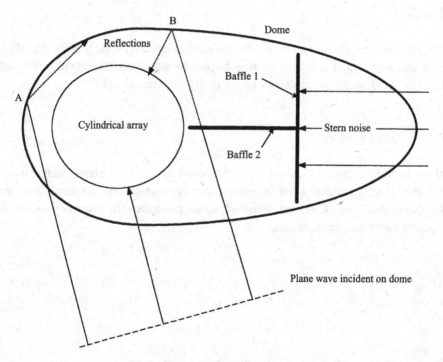

Figure 2.13   Baffled dome: plan view

The elongated dome meets the hydrodynamic requirements. The array, however, can only be close to the dome over about half its circumference. Baffle 1 reduces the stern noise. Performance astern (known in sonar as the non-operational bearings) will inevitably be impaired and this determines the position of baffle 1. It cannot be too close to the array without affecting too many non-astern beams.

The resultant gap between the array and baffle 1 introduces unwanted reflections. Therefore baffle 2 is introduced to eliminate the more significant reflections. Reflection A, in a region where the array is close to the dome, is harmless but reflection B (without the baffle) would contribute energy to several incorrect beams.

2.22 Problems

2.1 What are the beamwidths to the half-power points of a line array of length 10 m and frequency 4 kHz (i) normal (broadside) to the array axis and (ii) at a steer of 60°?

2.2 A line array has a length of 6λ. What is the level of the fourth-order sidelobes (in dB relative to the main lobe) and where do their peaks occur in the beam pattern?

2.3 What is the DI of a baffled planar array of length 5 m and height 2 m at 8 kHz? The elements are spaced $\lambda/2$ at 10 kHz. First calculate the DI using the numbers of elements and then compare with the result using the dimensions and wavelength.

2.4 What is the DI of a baffled cylindrical array of height 2 m and diameter 3 m at 5 kHz? How much would the source level of the array be reduced by halving its height?

3

Propagation of Sound in the Sea

3.1 Propagation Loss

Propagation loss (PL) is a quantitative measure of the reduction in sound intensity between the source and a distant receiver. If I_0 is the intensity of the source referred to a point one metre from its acoustic centre and I_r is the intensity at the receiver, then the propagation loss between source and receiver is

$$PL = 10 \log \left(\frac{I_0}{I_r} \right) \quad (\text{dB})$$

3.2 Losses

A first approach to quantifying the likely PL is to consider it as the sum of a *spreading loss* and a loss due to *absorption*. Other losses – not functions of range – include scattering and refraction, and these will be considered later on.

3.3   Spreading Losses

The spherical spreading law

Refer to Figure 3.1(a). When the source is located in an unbounded and lossless medium, the power is radiated equally in all directions and the total power, P, crossing spheres of increasing radii surrounding the source does not change with range. Therefore, since power $=$ intensity $\times$ area,

$$P = 4\pi r_1^2 I_1 = 4\pi r_2^2 I_2 = \cdots = 4\pi r^2 I_r$$

and if $r_1 = 1$ m, the PL to range r is

$$PL = 10\log\left(\frac{I_1}{I_r}\right) = 10\log r^2$$

or in logarithmic form,

$$\boxed{PL = 20\log r}$$

where r is in metres.

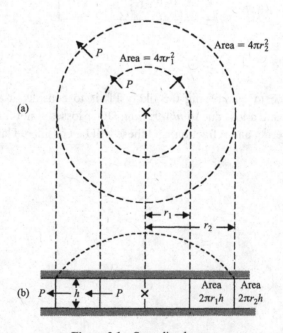

Figure 3.1   Spreading losses

The cylindrical spreading law

Refer to Figure 3.1(b). When the source is bounded by parallel planes separated by h metres, the power – again unchanging – crossing cylindrical surfaces of increasing radii surrounding the source, is given by

$$P = 2\pi r_1 h_1 I_1 = 2\pi r_2 h_2 I_2 = \cdots = 2\pi r h I_r$$

and if $r_1 = 1$ m, the PL to range r is

$$PL = 10\log\left(\frac{I_1}{I_r}\right) = 10\log r$$

or in logarithmic form,

$$\boxed{PL = 10\log r}$$

where r is in metres.

3.4   Absorption Losses

When a sound wave travels through sea water, absorption losses occur through two principal mechanisms:

- *Viscosity*: losses due to viscosity are present in fresh water and salt water. This contribution is proportional to the square of the frequency and accounts for the straight line for fresh water when α is plotted on a log-log graph like Figure 3.2.

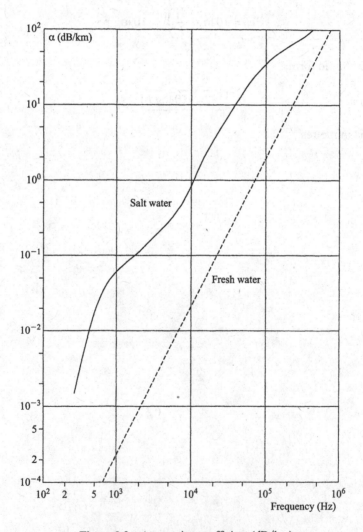

Figure 3.2   Attenuation coefficient (dB/km)

- *Molecular relaxation*: losses due to molecular relaxation are only present in salt water. The mechanism is a reduction of molecules to ions induced by the pressure of the sound. At very high frequencies (greater than about 500 kHz) pressure changes are too rapid for the relaxation to take place and therefore no energy is absorbed. Magnesium sulphate relaxation is dominant over the frequency range 2 to 500 kHz. Below 2 kHz boric acid relaxation contributes to the losses.

Extensive measurements of these losses have been made and several empirical formulae exist which relate them to frequency, depth (pressure) and salinity. The total loss is given as an attenuation coefficient, α, in dB/ km.

The *attenuation coefficient*, α, increases rapidly with frequency and changes with temperature. It also varies with depth and salinity, but less strongly. Figure 3.2 plots the variation of α with frequency at a temperature of 10°C and a salinity of 35 parts per thousand (ppt), based on the formulae of Francois and Harrison.

An approximation for α, useful between 0.5 kHz and 100 kHz, in 'standard' sea water is given by

$$\boxed{\alpha = 0.05f^{1.4}}$$

Table 3.1 is *not* based on the approximation; it offers a selection of values for α that will be useful in performance comparisons and in assessing the effects of frequency changes on a design. Linear interpolation will be sufficiently accurate to determine α at other frequencies.

Table 3.1 Values for α: not based on the approximation

T (°C)	α (dB/km)									
	0.5 kHz	1 kHz	2 kHz	5 kHz	10 kHz	20 kHz	50 kHz	100 kHz	200 kHz	500 kHz
5	0.02	0.06	0.14	0.33	1.00	3.80	15	30	55	120
10	0.02	0.06	0.14	0.29	0.82	3.30	16	35	60	125
15	0.02	0.06	0.14	0.26	0.68	2.80	17	40	65	130

3.5   Spherical Spreading and Absorption

The sum of spherical spreading and absorption losses is a useful working rule for
initial design and performance comparisons:

$$PL = 20 \log r + \alpha r \times 10^{-3} \quad \text{(dB)}$$

A remarkably good fit to measured propagation losses and to more complex
propagation models is frequently obtained. It is generally somewhat pessimistic,
as might be expected since there is usually some trapping which prevents spherical
spreading except at short ranges. Some workers use $15 \log r$ as a compromise
between spherical and cylindrical spreading.

Figure 3.3 plots spherical spreading loss alone and with the addition of

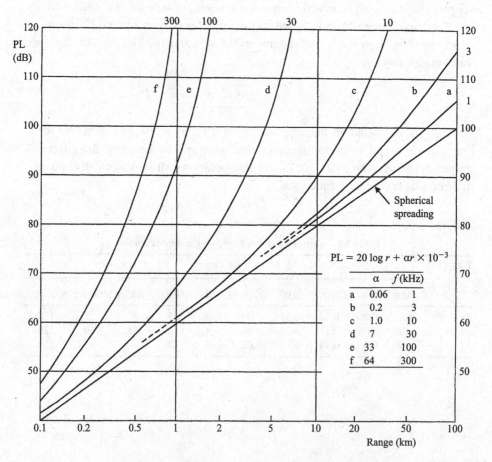

Figure 3.3   Propagation loss curves: effect of absorption

absorption at selected frequencies. Note that absorption is insignificant below 1 kHz, even at 100 km, whereas at 100 kHz absorption is prohibitive even at 2 km. It is interesting to compare sound and electromagnetic radiation losses in the sea:

- Sound   at 30 kHz, $\alpha = 5$ dB/km

- EM. wave   at 30 kHz, $\alpha_{em} = 7500$ dB/km

$$(\alpha_{em} = 1.4 \times 10^3 f^{1/2} \text{ dB/km})$$

This severe attenuation – even at low frequencies (low for electromagnetic radiation) – makes an 'underwater radar' no match for an acoustic (sonar) system in the highly conductive oceans.

3.6   Propagation in the Real Ocean

Refraction, scattering and the presence of the ocean boundaries – surface and bottom – ensure that, except at very short ranges, free-field conditions never exist in the real ocean and much effort is still being expended on developing reliable models which take account of this.

The speed of sound in the sea – its absolute value and, more importantly, its variation with depth – is fundamental to all models. A knowledge of the sound speed profile (SSP) can help the sonar designer and operator to choose the appropriate sonar propagation mode: a towed array or a variable depth sonar (VDS) may be placed at an appropriate depth to avoid shadow zones, or the beams of a hull-mounted sonar may be depressed to exploit bottom bounce or convergence zone (CZ) paths.

3.7   The Speed of Sound

The speed of sound in the sea depends on the temperature, pressure (depth) and salinity. A variety of empirical formulae exist for its calculation; here is one due to Leroy:

$$c = 1492.9 + 3(t - 10) - 6 \times 10^{-3}(t - 10)^2 - 4 \times 10^{-2}(t - 18)^2$$

$$+ 1.2(s - 35) - 10^{-2}(t - 18)(s - 35) + h/61$$

where

c = speed of sound (m/s)

t = temperature (°C)

s = salinity (ppt)

h = depth (m)

The speed of sound at 10°C, at zero depth, and for a salinity of 35 ppt is 1490 m/s. Here are some approximate coefficients for sound speed valid for use with this 'standard' sound speed:

- Temperature $\Delta c/\Delta t = +3.4$ m/s per °C
- Salinity $\Delta c/\Delta s = +1.2$ m/s per ppt
- Pressure (depth) $\Delta c/\Delta h = +17$ m/s per 1000 m

3.8   Sound Speed Profiles

Sound speed profiles (SSPs) are graphs of the speed of sound in the sea against depth. The SSP depends on the location, the season, the time of day and the weather. In most locations the salinity can be considered constant at 35 ppt, but some environments have a different salinity. Near estuaries the salinity is highly variable; in the Arctic melting ice means lower salinity near the surface; and in parts of the Baltic very low salinities are present at all depths.

Low salinity also means very low attenuation coefficients and therefore reduced propagation losses. A sonar designed for operation solely in the Baltic could, with useful reductions in size and cost, use higher frequencies than required for normal salinity.

It is the temperature of the sea as a function of depth that is the most variable and the most difficult to determine. It is usually measured by a *bathythermograph*, which may be deployed from vessels or aircraft, and typically has an accuracy or resolution of about 0.25°C.

The speed of sound increases with temperature and with depth. When the water near the surface is warmer than at greater depths, there are two opposing tendencies as the depth increases:

- The speed of sound decreases with decreasing temperature

- The speed of sound increases with increasing pressure

The result of these opposing tendencies is to produce SSPs which vary widely within the first few hundred metres of depth, and these SSPs are further complicated by diurnal changes as well as mixing of the surface layer by wind and waves.

Figure 3.4 shows a typical deep sea SSP divided into four principal layers:

- *Surface layer (duct)*: a layer of isothermal water mixed by the action of wind on the surface of the sea. Sound tends to be trapped in this layer by surface reflections and upward refractions.

- *Seasonal thermocline*: temperature decreases with depth. During summer and autumn the thermocline is strong and identifiable. During winter and spring it is weak and merges with the surface layer.

- *Main thermocline*: little affected by the seasons. This is where the main increase in temperature over the cold depths of the sea occurs. Although the pressure

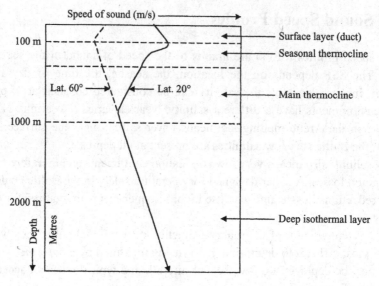

Figure 3.4 Typical deep sea SSP

increases with depth, the net effect of temperature and pressure changes is to reduce the sound speed through this layer.

- *Deep isothermal layer*: constant temperature of about 4°C right to the bottom. The speed of sound increases with increasing pressure. At high latitudes the layer extends closer to the sea surface, and in the Arctic it may completely eliminate the other layers. The tendency is shown in the dashed curve of Figure 3.4 for a latitude of about 60°.

3.9 Deep Sound Channel

Between the negative gradient of the main thermocline and the positive gradient of the deep layer there is a *sound speed minimum*, where sound tends to be focused by refraction. The depth at which this focusing occurs is known as the *deep sound channel* (DSC). To exploit this channel, the source is placed close to the minimum (which may be only a few hundred metres at high latitudes) and, because the spreading is cylindrical, very long-range propagation is possible. Best results are achieved when the receiver is close to the axis of the channel. Figure 3.5 shows a simplified sound ray plot for a source placed at a depth close to the sound speed minimum and with a vertical beamwidth of about 20°.

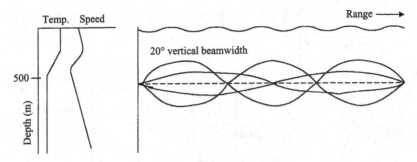

Figure 3.5   Deep sound channel propagation

3.10   Reliable Acoustic Path

Placing the source deep in the sea (at least 1000 m) can improve detection of shallow targets, i.e. targets at typical submarine diving depths (Figure 3.6). The path is known as 'reliable' because it is insensitive to highly variable surface effects and bottom losses. Conditions for reliable acoustic path (RAP) exist when the source is placed at a depth, the *critical depth*, where the sound speed is equal to the sound speed at the surface (Figure 3.7).

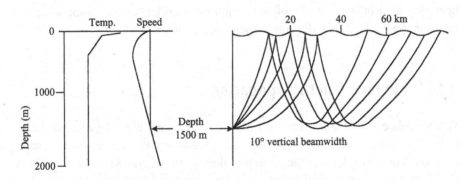

Figure 3.6   Deep source: reliable acoustic path

Note that, particularly at high latitudes, the DSC may be close to the RAP critical depth. A capability to steer the sonar beam vertically by, say, 5° (assuming a 10° vertical beamwidth) as well as a capability to deploy the sonar at variable depths will assist in reducing the shadow zones that would otherwise exist.

The great depths at which it is necessary to deploy both source and receiver – at

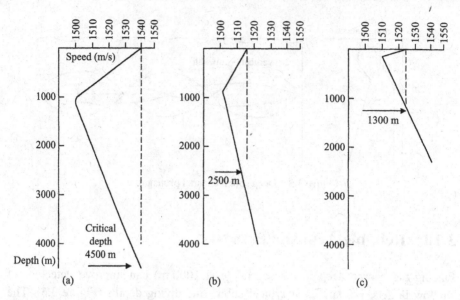

Figure 3.7   Critical depths for RAP: (a) tropical, latitude 20°; (b) temperate, latitude 50°; (c) Mediterranean

least 500 m for DSC mode and 1000 m for RAP mode – impose severe engineering and mobility constraints on their use. The design of elements capable of operating at depths in excess of, say, 1000 m is difficult if not impossible and losses in the long cables will limit the achievable source levels.

3.11   Surface Duct Propagation

When surface winds and waves mix the shallow layers of the sea to produce a nearly isothermal layer, the pressure effect dominates and the sound speed increases down to a depth – the bottom of the layer – where the temperature starts to fall and the sound speed begins to reduce until the minimum of the DSC is reached. This isothermal layer, the *surface duct*, can be as small as 5 m and as great as 200 m. Typically, ducts of 50–100 m are common in the colder waters of the world.

Figure 3.8 shows the effect on sound transmitted from a source within the duct. Rays that are projected close to horizontal are refracted upwards and undergo multiple surface reflections. On the other hand, rays which penetrate the layer are refracted downwards at first, thus producing a zone – known as a shadow zone – where hardly any sound energy penetrates. Targets within the shadow zone, i.e.,

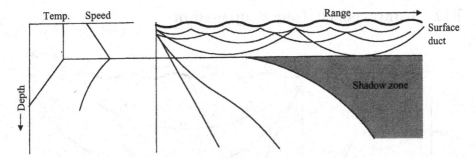

Figure 3.8 Surface duct propagation: shallow source

below the layer, are therefore difficult to detect. As with all propagation in the sea, no mode is perfectly described by a simplified ray trace. The shadow zone is an area where the sound intensity is greatly reduced and the transition from the surface duct is not abrupt.

Increasing the depth of the source so that it is below the layer (Figure 3.9) has the effect of increasing the range of the start of the shadow zone, but it may then extend into the duct. As with DSC and RAP modes, the ability to manoeuvre in depth – a variable depth sonar (VDS) deployed from a moving surface ship or a hovering helicopter – offers significant operational advantages.

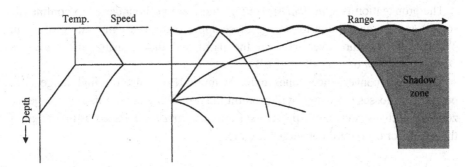

Figure 3.9 Surface duct propagation: deeper source

3.12   Convergence Zone Propagation

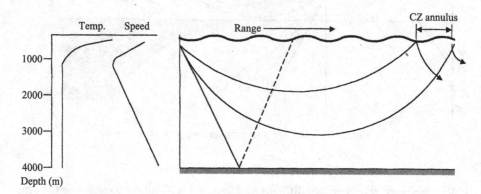

Figure 3.10   Convergence zone propagation

Rays projected at steeper angles of depression, 5° or more, are bent downwards at first – producing a shadow zone. At greater depths the pressure bends these rays upwards to form annuli of high intensity (Figure 3.10); each annulus is known as a *convergence zone* (CZ). The water must be deep enough for upwards refraction to prevent the rays from hitting the bottom. Typically the water depth must be in excess of 3000 metres. Depending on the bottom depth, the first convergence zone will occur at around 30–50 km and it will be 3–5 km wide.

The propagation is *spherical spreading* (there are no boundaries to confine the sound) and the absorption losses are those appropriate to a temperature of about 4°C (the temperature over most of the path) and the focusing effect gives a *convergence gain* of typically 3–6 dB.

Successive convergence zones exist at multiples of the original range. The propagation losses (two-way) will prohibit the use of any but the first convergence zone for active systems, but passive systems can make detections at the range of the second or even third convergence zone.

3.13 Bottom Bounce Propagation

Propagation is possible by using bottom reflections (Figure 3.11). The sonar beam is now deliberately trained downwards at relatively steep angles. The effectiveness of the mode is determined by the nature of the bottom, whether it is absorptive or reflective, and how the bottom loss varies with angle of incidence. As with convergence zones, there exists a range annulus, which varies with depression angle of the sonar beam, and at small angles of incidence this annulus can be very wide. No focusing gain exists and the bottom reflection loss is typically between 10 and 20 dB. Therefore the mode is very demanding in projector power and array size (because of the lower frequencies necessary to limit absorption over the long-range paths).

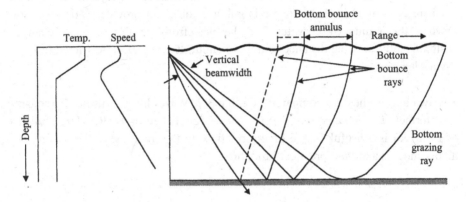

Figure 3.11   Bottom bounce propagation

An active bottom bounce (BB) system will be large, very demanding in ship's power supplies and transmitters and requiring a large site for the installation of its arrays. Successive range annuli exist at multiples of the first annulus. The propagation losses (two-way) will prohibit the use of any but the first for active systems, but passive systems can make detections at the ranges of the second annulus. The range annuli for BB mode are significantly wider than for the focused CZ mode. There must always be a shadow zone out to some minimum range, at least 10 km, dependent on depth and the allowable angle of depression of the sonar beam (allowable from the viewpoint of bottom loss).

3.14   Propagation Loss Models

The spreading, refraction and reflection of sound in the sea have so far been discussed in a qualitative manner only, except with very simple – but often adequate – quantitative expressions. The propagation of sound in an elastic medium is described mathematically by solving the *wave equation* using the appropriate boundary and medium conditions for a particular environment.

There are two approaches to solving this equation:

- *Wave theory*: the propagation is described in terms of characteristic functions called normal modes, each of which is a solution to the equation. The modes are combined additively to satisfy the required boundary and medium conditions.

- *Ray theory*: this postulates wavefronts and the existence of rays which indicate where the sound from the source is going. It does not provide a good solution when the radius of curvature of the wavefront or the pressure changes significantly over the distance of a wavelength. It is therefore restricted to short wavelengths.

The two approaches are compared in Table 3.2. Wave theory will not be considered further. The *Sonar Modelling Handbook* (published by DERA, UK Ministry of Defence) is a useful source of information on the many wave theory and ray theory models and their practical limitations.

Table 3.2   Wave theory versus ray theory

Wave theory	Ray theory
Formal, complete solution	Rays easily drawn
Difficult to interpret	Easy to visualize sound distribution
Real boundary conditions difficult to handle	Real boundary conditions easy to insert
Valid at all frequencies; in practice useful at low frequencies where ray theory fails	Valid only at 'high' frequencies (> 200 Hz?)
Computer program essential	Computer program normally used, but rays can be drawn manually using Snell's law

3.15 Ray Theory and the Hodgson Model

Ray theory will be discussed in terms of the Hodgson model. This model is used operationally by the Royal Navy and is applicable at frequencies above, say, 200 Hz. Sonar parameters that may be specified are frequency, depth, and beam minimum and maximum (vertical) angles.

The sea surface is modelled as a reflector at which rays undergo an attenuation per bounce dependent on frequency and surface roughness (sea state, wind speed or wave height may be input). Bottom losses are calculated using a set of bottom loss curves; the loss per bounce depends on frequency and grazing angle. Absorption losses are accurately modelled as a function of frequency, temperature, depth and salinity and are continuously changed with range for each ray.

The sound speed field is calculated by interpolating the SSPs in depth and range then the ray paths are calculated using Snell's law, after splitting the environment into a series of boxes within which the sound speed varies linearly with depth (Figure 3.12). This is how a range-dependent solution is derived.

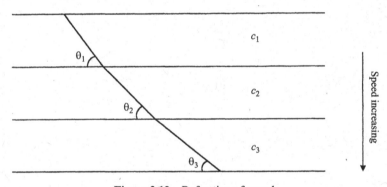

Figure 3.12   Refraction of sound

Snell's law

Snell's law is fundamental to ray theory models and describes the refraction of sound rays in a medium where the speed of sound is changing. When the speed of sound varies continuously with depth, the medium can be considered as a number of thin layers of constant but different sound speeds within them. Snell's law is applied to the boundaries of the layers, and the sound ray is seen to be curved:

$$\frac{\cos \theta_1}{c_1} = \frac{\cos \theta_2}{c_2} = \frac{\cos \theta_3}{c_3}$$

If speed continues increasing with depth, the ray will become horizontal and $\cos \theta = 1$. The speed of sound where the ray is horizontal, c_v, is known as the *vertex speed*, and

$$\frac{\cos \theta}{c} = \frac{1}{c_v}$$

This equation makes it possible to generate the path of a ray through the layers into which the velocity profile has been divided.

3.16 Hodgson Example

Propagation losses in a central Mediterranean location are shown for February (Figure 3.13) and August (Figure 3.14). The frequency is 5 kHz and the sonar vertical beamwidth is 0–15°. The source depth is 5 m.

February

The ray trace shows shadow zones between about 8 and 25 km and this is evident in the PL curves for the two receiver depths of 50 and 200 m. Note that the simple law of spherical spreading plus absorption follows the model predictions quite accurately, particularly if the shadow zones are excluded. The reduced losses centred on about 30 km are due to a CZ gain.

August

Severe downward refraction occurs near the surface, resulting in a shadow zone from about 2 km until the CZ returns, centred on about 40 km. Note the first BB returns and the second CZ returns (at greater than 110 dB each way, the losses will be too great for an active system but passive will sometimes be possible).

Figures 3.13, 3.14 and 3.15 have been produced using the *Wader–Hodgson system*. Intellectual property rights and copyright to the Hodgson Acoustic Propagation Loss model are owned by the United Kingdom Secretary of State for Defence. Intellectual property rights and copyright to the Wader Global Ocean Information System are owned by Ocean Acoustic Developments.

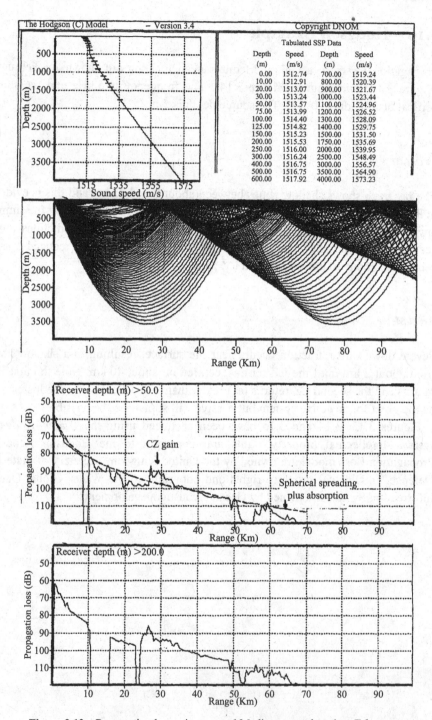

Figure 3.13   Propagation losses in a central Mediterranean location: February

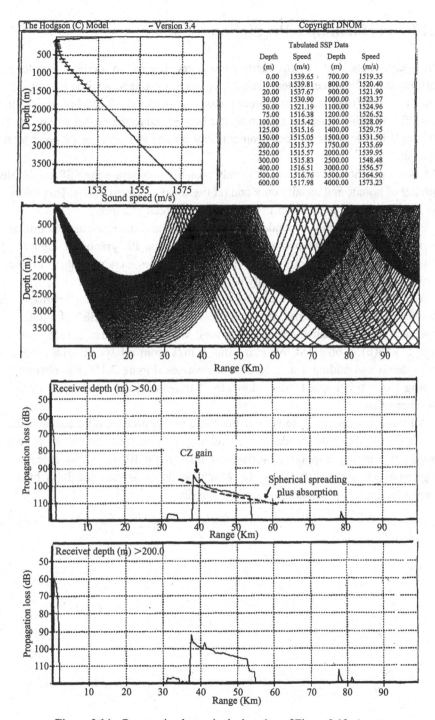

Figure 3.14   Propagation losses in the location of Figure 3.13: August

3.17 Performance Prediction

Propagation loss is one of the many parameters required to predict the detection performance of a sonar system. Parameters such as source level (SL) and directivity index (DI) are fairly accurately known, to within say 2 dB, but propagation loss (PL) is highly variable, and no matter how accurate the model and no matter how well it represents the environment, any results can only be as good as the input data.

Under operational conditions, it is only possible to measure the SSP at a limited number of points, and usually only one (at the sonar platform itself). This SSP will not apply precisely to all water columns between sonar and target.

PL graphs are highly dependent on the SSPs, which themselves rely primarily on measurements of temperature against depth. The PL graphs in Figure 3.15 illustrate the differences for quite small inaccuracies in temperature measurements, inaccuracies typical of bathythermographs

All the graphs are for an area in the North Arabian Sea during February. The 'standard' SSP for February yields Figure 3.15(a). When this SSP is modified to represent possible bathythermograph errors, we obtain Figures 3.15(b) and (c). Figure 3.15(b) is obtained by subtracting 1 m/s from all speed values down to 75 m depth and adding 1 m/s to all other values. Figure 3.15(c) is obtained by adding 1 m/s to all speed values down to 75 m depth and subtracting 1 m/s from all other values.

All three graphs are quite similar but there are significant differences in detail which should illustrate the dangers of relying implicitly on the results from any propagation model. Suppose a propagation loss of 100 dB is the allowable loss for a given sonar system (the loss for which detection is just possible) then the corresponding range limits differ quite markedly.

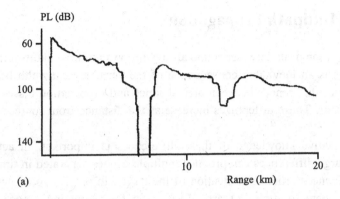

(a)

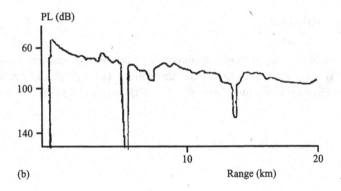

(b)

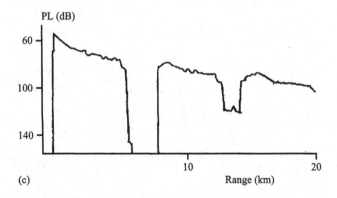

(c)

Figure 3.15 Propagation loss graphs for the North Arabian Sea: (a) standard SSP, (b) modified SSP, (c) modified SSP

3.18   Multipath Propagation

The propagation of underwater sound always follows multiple paths in the vertical plane. This is an inevitable consequence of the vertical beamwidth being finite, and it causes differences between arrival times and propagation losses along the possible paths. These differences increase as the distance from source to receiver increases.

A quantitative knowledge of these differences is important. In active sonar systems, large differences can produce multiple echoes separated in time (range); small differences produce elongation of the target. In passive sonar systems, the differences between times of arrival are used for estimating range. Multipath propagation is considered further in Chapters 8 and 9.

3.19   Problem

3.1 What are the detection ranges achieved for a one-way propagation loss of 80 dB assuming (i) spherical spreading plus absorption, (ii) cylindrical spreading plus absorption? Give solutions for 5 kHz ($\alpha = 0.3$ dB/km) and 20 kHz ($\alpha = 3$ dB/km).

4

Target Strength

4.1 Definition

Target strength (TS) refers to the echo returned by an underwater target: submarines, surface ships, torpedoes, mines, fish. It is defined as the log of the ratio, in dB, of the *reflected intensity* referred to 1 m from the acoustic centre of the target, to the *incident intensity*:

$$TS = 10 \log \left(\frac{I_r}{I_i} \right)$$

The use of an arbitrary reference distance of 1 m gives many underwater targets *positive* values of TS. This does not imply that more sound is reflected from the target than is incident on it. The acoustic energy appears to come from a hypothetical point source within the scattering surface of the target, and for a large target such as a submarine, this point may be some metres from the surface and *within* the target.

4.2 Formulae

The correct value of TS to be used in the sonar equations should be chosen carefully. In practice, TS is calculated using either the peak pressures of the incident and reflected pulses or their total integrated energies. Here are the resultant parameters:

• Peak TS

$$\text{Peak TS} = 20 \log(p_r / p_i)$$

• Integrated TS

$$\text{Integrated TS} = 10 \log \left(\int_0^{T_e} p_r^2(t) \mathrm{d}t \bigg/ \int_0^{T_p} p_i^2(t) \mathrm{d}t \right)$$

where p_i and p_r are the peak pressures of the incident and reflected pulses; $p_i(t)$ and $p_r(t)$ are the time functions of the pulses; T_p is the duration of the incident pulse and T_e is the time extent of the target.

4.3 Measurement

It is simplest to measure the peak pressures of the incident and reflected pulses. Therefore peak TS is usually determined regardless of the method of measurement and is the parameter normally used in the active sonar equations.

TS is defined at 1 m from the acoustic centre of a target. It is clearly impractical and often impossible to make measurements at this distance. Therefore measurements are made at greater distances and reduced to the definition range.

A method well suited to the measurement of the TS of small objects such as mines is to compare the echo levels from the target with the level from a reference target such as a sphere. For larger targets such as submarines and torpedoes, which because of their size must be measured at longer range, say 1000 m, the necessarily large size of the reference target (a sphere of TS = 0 dB has a diameter of 4 m), the reference target may be replaced by a calibrated transponder or an alternative method may be used.

Most TS determinations, particularly for these larger targets, have been made by measuring the peak pressure of the reflected pulse at long range and then reducing it to its 1 m value. The appropriate sonar equation is

$$\text{EL} = \text{SL} - 2\text{PL} + \text{TS}$$

The echo level (EL) and the source level (SL) are measured, the propagation loss (PL) is either calculated, assuming spherical spreading plus adsorption, or measured using a calibrated hydrophone at the target, and the equation is solved for TS.

4.4 Dependence on Pulse Type and Duration

Integrated target strength (ITS) does not change with duration or type of pulse, but fluctuations will occur from ping to ping, given the effect on TS of even small aspect changes. These fluctuations will be averaged out so that data from the same target should yield very similar ITS values irrespective of pulse type and duration.

Peak target strength (PTS), on the other hand, is very dependent on pulse type and duration. It increases with pulse duration, reaching a constant value when the pulse duration is comparable with the time extent of the target – in practice at least 100 ms for a submarine and as short as 1 ms for a mine. For such long pulses (long relative to the target dimensions) PTS is approximately equal to ITS.

Modern active sonars frequently use long CW (continuous wave) and long FM (frequency modulated) pulses. It is important therefore to know which TS value to use for these pulses:

- *Long CW*: the small bandwidth and thus low resolution of this pulse type ensures the PTS is approximately equal to the ITS.

- *Long FM*: the large bandwidth and thus high resolution of this pulse type means that the effective pulse length is short, of the order of a few milliseconds for an antisubmarine pulse, and therefore the correct TS value will be given by the PTS as measured by an equivalent short pulse.

4.5 TS of a Sphere

The meaning of target strength can be demonstrated by calculating the TS of a sphere. Let a large sphere – large compared with a wavelength – be insonified by a plane wave of intensity I_i. If the sphere has radius a metres, the energy intercepted by it from the incident sound is $\pi a^2 I_i$; where πa^2 is the cross-section of the sphere, the *scattering cross-section*.

Assuming the sphere reflects this energy isotropically, the intensity of the reflected wave at a distance r metres from the centre of the sphere will be $I_r = \pi a^2 I_i / 4\pi r^2$. At the reference distance of 1 m this reduces to $I_r / I_i = a^2 / 4$ and the target strength of the sphere is given by

$$TS = 10 \log\left(\frac{a^2}{4}\right)$$

Therefore a sphere of radius 2 m has TS = 0 dB.

4.6 TS of Some Simple Shapes

Formulae for target strengths of other simple forms have been derived (Table 4.1) and these may be used to estimate target strengths of practical targets. The minimum dimension of the shape should be large compared to a wavelength (say at least 5λ), although a useful *indication* of target strength may still be obtained even when the minimum dimension is only 2λ.

Table 4.1   Target strengths of some simple shapes

Shape	TS, dB	Incidence	Notes
Sphere	$10\log(a^2/4)$	Any	a is radius
Convex surface	$10\log(a_1a_2/4)$	Normal to surface	a_1 and a_2 are principal radii
Plate of any shape	$10\log(A/\lambda)^2$	Normal	A is area
Rectangular plate	$10\log(ab/\lambda)^2$	Normal	a and b are
	$10\log(ab/\lambda)^2 + 20\log(x^{-1}\sin x) + 20\log(\cos\theta)$ where $x = (2\pi a/\lambda)\sin\theta$	θ to normal	sides $a \geqslant b$
Circular plate	$10\log(\pi a^2/\lambda)^2$	Normal	a is radius
Cylinder	$10\log(aL^2/2\lambda)$	Normal	a is radius L
	$10\log(aL^2/2\lambda) + 20\log(x^{-1}\sin x) + 20\log(\cos\theta)$ where $x = (2\pi L/\lambda)\sin\theta$	θ to normal	is length

Note that the scattering cross-section is known in radar as the *radar cross-section* (RCS or σ). Expressions for the RCS of simple shapes are prevalent in the radar literature. If helpful, these expressions can be translated into sonar TS by *dividing them by 4π*. The target strengths of certain shapes at an angle θ to the normal may be determined with the aid of Figure 4.1, which plots $20\log(x^{-1}\sin x)$ dB against $(L/\lambda)\sin\theta$.

Example 4.1

A cylinder has length $L = 5$ m, radius $a = 1$ m and $\lambda = 0.2$ m. What is the TS at an angle of $2°$ to the normal?

$$TS = 10\log(aL^2/2\lambda) + 20\log(x^{-1}\sin x) + 20\log(\cos\theta)$$

$$(L/\lambda)\sin\theta = 0.87$$

and from the plot we obtain

$$20\log(x^{-1}\sin x) = -16 \text{ dB}$$

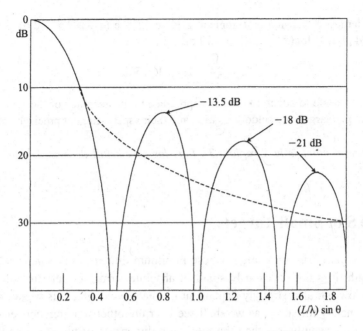

Figure 4.1 Plot of $20\log(x^{-1}\sin x)$ against $(L/\lambda)\sin\theta$

Therefore TS $= 18 - 16 + 0 = 2$ dB. At the first null, $(L/\lambda)\sin\theta = 0.5$, therefore $\theta = 1.15°$ and TS is theoretically $-\infty$.

In practice, perfect nulls do not exist; the cylinder will not be perfectly uniform and the incident pulse will have a finite bandwidth, i.e., the wavelength will vary during the pulse. Broadband pulses (FM or short CW) will have significant percentage bandwidths, perhaps 5 per cent and often greater. Therefore, particularly for broadband pulses, the nulls will be blurred and the peaks broadened. Narrowband pulses (long CW) will have less effect. The dashed curve is a realistic plot avoiding the extreme nulls and peaks. If it is used for this example, we obtain the following results:

- For $\theta = 1.15°$ TS $= 18 - 18 + 0 = 0$ dB
- For $\theta = 2°$ TS $= 18 - 24 + 0 = -6$ dB

Example 4.2
A rectangle has sides $a = 5$ m, $b = 3$ m and $\lambda = 1$ m. What is the TS at an angle of $7°$ to the normal?

$$TS = 10\log(ab/\lambda)^2 + 20\log(x^{-1}\sin x) + 20\log(\cos\theta)$$

Now the largest dimension is a, replacing L, so we use $(a/\lambda)\sin\theta = 0.61$ and, from the dashed plot, $20\log(x^{-1}\sin x) = -19$ dB. Therefore

$$TS = 24 - 19 + 0 = 5 \text{ dB}$$

It would be unsafe to attach too much importance to the accuracy of these theoretical TS values; real targets are seldom regular, and their aspects are not precisely known.

The important point is how rapidly TS falls for very small departures from normal incidence.

4.7 TS of Small Targets

A small target is simply a target whose minimum dimension is *much less than* a wavelength; it is not necessarily small in absolute terms. Target strength is then not only low but also strongly dependent on the wavelength; it is proportional to λ^{-4}. Fortunately for sonar, as we shall see later on, other requirements ensure the wavelength is usually less than the minimum dimension of the target so that this region is avoided.

Resonance is possible when a target has one or more dimensions *close to* λ; TS can then vary wildly with λ and may be greater or smaller than the values indicated above. This resonance effect could be exploited by choice of frequency to *improve the detection* of sonar targets at some possible cost to classification.

Suppose a submarine to have a minimum dimension of 8 m. To exploit a resonance for this dimension, $f = c/\lambda = 188$ Hz. To use other resonances, even lower frequencies may be needed and sufficient bandwidth for classification will not be possible. There are formidable obstacles to this approach: frequency diversity would be needed to achieve resonances, and at the low frequencies necessary for submarine resonances, both the transmit and receive arrays would have to be very large.

4.8 Mine Target Strength

Mines may often be approximated by a sphere or a cylinder with hemispherical ends. The TS of a finite cylinder is given by

$$TS = 10\log\left[\frac{aL^2}{2\lambda}\left(\frac{\sin x}{x}\right)^2 \cos^2\theta\right]$$

where

a = radius

L = length

$x = \dfrac{2\pi L}{\lambda} \sin \theta$

On the beam (normal to the axis) this reduces to

$$TS = 10 \log \left(\frac{aL^2}{2\lambda} \right)$$

Suppose the mine is represented by a cylinder of length 2 m with hemispherical ends of radius 0.15 m and with $\lambda = 0.15$ (10 kHz).

- TS on the beam = $10 \log\{(0.15 \times 4)/(2 \times 0.15)\} = 3$ dB
- TS on the ends = $10 \log(0.15^2/4) = -22.5$ dB

As we have seen, the TS falls rapidly away from its value on the beam, but because of the hemispherical ends it does not fall below -22.5 dB.

4.9 Torpedo Target Strength

A torpedo is basically cylindrical with a flat or rounded nose, and the same formulae may be used to estimate its TS. Suppose the torpedo has length $L = 5$ m and radius $a = 0.26$ m, and the frequency is 10 kHz ($\lambda = 0.15$):

- TS on the beam = $10 \log\{(0.26 \times 25)/(2 \times 0.15)\} = 13$ dB
- TS on the (rounded) nose = $10 \log(0.26^2/4) = -18$ dB

Again, the TS falls rapidly away from normal aspect (on the beam), but because of the nose, it will not fall below -18 dB, except perhaps astern.

Some torpedo designs have flat noses to reduce flow noise and therefore improve homing sonar performance. TS on the (flat, circular) nose at *normal incidence* is

$$10 \log(\pi \times 0.26^2/0.15)^2 = 3 \text{ dB}$$

Once again, this high TS value will fall off rapidly away from normal and a *flat nose torpedo* will be virtually undetectable by an active sonar at any aspects other than close to beam or bow.

4.10   Submarine Echoes

Echoes from submarines include specular reflections from the outside profile of the submarine; reflections or scattering from structures on or behind the casing or pressure hull. The echoes are caused by a reflection from a surface normal to the incident wave and are called normal incidence specular reflections. All other reflections from the outside profile are at non-normal angles and directed away from the receiver.

At sonar search frequencies the water-backed casings and all free-flood areas, including the fin, are virtually transparent and therefore reflections from these outer surfaces are at low intensities. The pressure hull, being air-backed, provides good reflections and a significant echo.

At weapon frequencies (> 20 kHz) both casings and pressure hull are good reflectors and echoes can be expected from the external casings, pressure hull, fin, rudder, hydroplanes, stabilizers and propeller. Shadowing, due to target aspect, will affect the number of highlights.

4.11   Beam Aspect Target Strength

For submarines (and mines and torpedoes) specular reflection appears to be the dominant mechanism at beam aspect and the incident pulse is faithfully reproduced by the echo. If we take the dimensions of the uniform section of the cylindrical pressure hull of a submarine to be 50 m and its diameter 8 m, then using

$$TS = 10 \log \left(\frac{aL^2}{2\lambda} \right) \quad \text{(at normal incidence)}$$

at 5 kHz we have

$$TS = 10 \log \left(\frac{4 \times 2500}{2 \times 0.3} \right) = 42 \text{ dB}$$

The 50 m 'aperture', equivalent to about 150λ at 5 kHz, ensures that this broadside 'glint' only occurs at this intensity over a fraction of a degree. But since the hull tends to taper and is seldom exactly cylindrical, this large value is reduced and exists over greater angles (compare with the narrow beamwidths achieved by long towed arrays of large apertures). The fact that the TS varies rapidly with aspect helps to explain the extreme variability of practical measurements.

4.12 Bow Aspect Target Strength

If the bow were hemispherical in form and 8 m in diameter, then the TS would be 6 dB over the entire bow sector. The submarine designer can easily reduce this wide coverage in both azimuth and elevation by avoiding a hemispherical shape and angling surfaces so that any glints occur at harmless elevations.

4.13 Submarine Target Strengths

Submarine target strengths are clearly directly related to the size and construction of the class of submarine. Typical submarines are listed in Table 4.2 and have been chosen based on their actual or potential exports or use by non-NATO navies. Six R2 Mala, for example, have been delivered to Libya; they carry 250 kg of limpet mines.

Table 4.2 Some typical submarines

Name	Mass (tonnes)	Dimensions (m)	Country
		LARGE SUBMARINES	
KILO	3000	74 × 10 × 6.6	Russia
209	1300	56 × 6.2 × 5.5	Germany
RUBIS (nuclear)	2700	74 × 7.6 × 6.4	France
		SMALL SUBMARINES	
Mini-sub	150	27	Italy, Maritalia
Mini-sub	90	20 × 2 × 1.6	North Korea
3GST9	30	10	Italy, Maritalia
R2 Mala	1.4	4.9 × 1.4 × 1.3	Former Yugoslavia

A submarine may be double-hulled, a complete external casing over an inner pressure hull, or single-hulled with much of the pressure hull directly exposed to the incident sound. The double-hulled construction may have main support ribs that are internal or external to the pressure hull. If they are external, they will add many potential highlights distributed along the length of the hull. For the single-hulled vessel there may be an upper casing and support ribs with winches housed beneath. The vessel may have a keel. All submarines have a bridge fin or sail (except for the very smallest submarines), and because it is proportionately larger, it is a more important echo source in smaller submarines. It is possible to examine the details of each vessel and by theoretical analysis estimate the TS of a particular design. The scattering mechanisms are diverse and often highly dependent on frequency and aspect, therefore the sonar designer is unlikely to be helped by such analysis.

Practical measurements of TS result in similar dependencies together with measurement uncertainties. Many TS measurements are reported in the literature and it is often claimed that they lend support to a standard 'butterfly' pattern as shown in Figure 4.2.

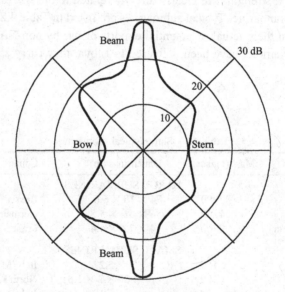

Figure 4.2   Aspect dependency of submarine target strength: 'butterfly' pattern

4.14   Towed Arrays

Towed arrays and cables are long, thin cylinders and echoes have been reported from them at sonar search frequencies. Maximum target strengths can theoretically be as high as 10 dB, but again *only very close to broadside and neglecting any array curvature*. Thin arrays will have lower maximum TS and this – combined with submarine tactics which generally attempt to avoid presenting themselves (and therefore their towed arrays) at beam aspect – implies that any contribution from a towed array either for detection or classification is unlikely to be useful operationally.

4.15   Target Strength Reduction

Careful attention to the *shape* and *orientation* of the external surfaces of a
submarine will reduce the target strength and ensure that glints only occur at
harmless angles, particularly in elevation. But note that homing torpedoes may
approach at angles far from the horizontal, therefore harmless angles for search
sonars may not be harmless for weapon sonars.

For areas of a submarine's external surface where only scattering returns are
likely, useful reductions in target strength are obtained by providing a coating with
a high transmission loss. Conventional 'decoupling' coatings can be effective
down to about 3 kHz.

Sound absorption is needed to reduce specular reflections at normal incidence.
This can be provided by a material 'thin' in terms of wavelength with limited
bandwidth, or a material 'thick' in terms of wavelength but relatively broadband.
Figure 4.3 shows the theoretical performance of a *thick-layer anechoic material*.

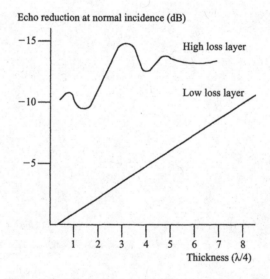

Figure 4.3   Echo reduction: thick layer anechoic material

To obtain a useful reduction in echo strength, the material needs to be at least
$\lambda/2$ thick. At weapon frequencies, say 30 kHz, $\lambda/2 = 50$ mm, which is practical
for large submarines. At 3 kHz, however, $\lambda/2 = 500$ mm and this is clearly
impractical.

A submarine made less vulnerable to weapons by a 50 mm coating – perhaps
achieving a useful reduction of between 10 and 15 dB at many aspects – might

only achieve a reduction of about 3 dB at typical sonar search frequencies (3–10 kHz) and no reduction at all at frequencies less than 3 kHz.

Thin tuned coatings

A prototype thin tuned coating was used in the Alberich design employed by the Germans during World War II. It consists of an inner layer of rubber perforated with a pattern of holes and covered by an outer layer of similar thickness (2 mm). Resonances occur and the attenuation is high over limited bands. At search sonar frequencies the resonant approach is essential, and in order to provide adequate transmission loss, the sound absorber may be combined with a decoupling base layer. The resonances depend on depth and temperature (Figure 4.4).

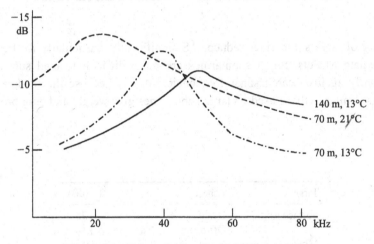

Figure 4.4 Echo reduction: typical Alberich design

4.16   Practical Values

The following TS values are suggested for use in the design of sonar systems and for performance calculations.

Submarines

Aspect	TS (dB)		
	Small	Large: clad	Large
Beam	5	10	25
Intermediate	3	8	15
Bow/stern	0	5	10

Cladding of large submarines reduces TS significantly, particularly for beam and intermediate aspects. Small submarines are more difficult to clad successfully, particularly against search sonar frequencies; this is because the thicknesses of coatings similar to those used on large submarines give weight and drag problems.

Others

Target	Aspect	TS (dB)
Surface ship	Beam	+25
	Off-beam	+15
Mines	Beam	0
	Off-beam	−10 to −25
Torpedoes	Random	−15
Towed array	Beam	0 (max)
Whale, 30 m	Dorsal	+5
Shark, 10 m	Dorsal	−4
Iceberg	Any	+10 (min)

4.17 Problems

4.1 What is the target strength of a sphere of radius 0.5 m? Above what frequency is the result likely to be reliable?

4.2 A cylindrical mine has a length 2 m and radius 0.5 m. The ends are hemispheres. What are the target strengths normal to the mine and end on? Calculate TS for frequencies of 10 kHz and 100 kHz.

4.3 What is the target strength of an irregular plate of area 4 m^2 and largest dimension 2 m at an angle of 5° to the normal and frequency 4 kHz? Use the practical dotted curve of Figure 4.1 and take L to be 2 m, the largest dimension.

5

Noise in Sonar Systems

5.1 Sources of Noise

Noise is the background against which sonars, active and passive, must detect signals from targets. For an active sonar, noise is augmented by reverberations from unwanted sources (Chapter 6) and the signal is an echo from the target. For a passive sonar, the signal is also noise – the radiated noise of the target.

This chapter will first discuss the sources of noise in sonar systems and then consider the total noise as a background to detection – the *self-noise* of a sonar system. There are three sources of noise to be considered:

- Thermal noise
- Noise from the sea (ambient noise)
- Vessel noise

5.2   Thermal Noise

In common with any electrical receiving system, a sonar receiver adds its own noise to the signals it receives. The designer must ensure that the noise introduced from this source is negligible compared with the noise coming from the sea itself. All the results established for the classical case of radio are valid for sonar when 'aerial' or 'antenna' is replaced by 'hydrophone'.

Any resistance, R, is the source of a *thermal noise EMF* (electromotive force) resulting from the thermal agitation of its electrons. The value of this EMF, e_n volts, is given by

$$e_n = \sqrt{4RkT\,\delta f}$$

where

$k =$ Boltzmann's constant $= 1.38 \times 10^{-23}$ J/K

$T =$ absolute temperature (K) of the resistance

$\delta f =$ bandwidth (Hz)

A practical formula, valid at sea temperatures, is

$$\boxed{e_n = 0.13\sqrt{R\delta f}}$$

where e_n is in microvolts, R is in kilohms and δf is in kilohertz.

A passive circuit produces a noise proportional to the resistive component of its equivalent impedance, even when this corresponds to a physical reality far removed from a simple resistance. For an *aerial*, a component of its impedance corresponds to the coupling with the surrounding space and to exchanges of energy with this space. This component is known as the *radiation resistance* of the aerial.

Similarly, for an underwater transducer, energy exchanges take place with the sea through the *motional resistance*, R_m, of the transducer, generating a noise EMF given by the above formula where R is replaced by R_m. Here the noise EMF does not come from a thermal agitation of electrons, but from the thermal agitation of the molecules of water producing pressure fluctuations at the face of the hydrophone.

This *thermal noise* is an absolute minimum and would only be observed in the absence of any other noise source, i.e.,

- If the water had no other agitation except thermal agitation and if it were completely isolated from any source of sound (a 'dead' sea)

- If the receivers were perfect and added no noise (the noise factor of the receiver is $NF_r = 0$ dB).

The thermal noise is given by

$$\boxed{N_{\text{thermal}} = -15 + 20 \log f}$$

where N_{thermal} is in dB re 1 μPa, and f is in kilohertz.

5.3   Noise from the Sea

The thermal noise of the sea can only be the dominant background to a sonar at high frequencies – at least 30 kHz and for all practical purposes at least 100 kHz, where it equals the ambient noise expected from sea state 2 (SS2).

When the sea is not 'dead' (i.e., perfectly isolated from all sources of sound and only subject to thermal agitation), even though it may appear perfectly calm, it is subject to an agitation much greater than the thermal noise, particularly at the lower frequencies (less than 30 kHz). Figure 5.1 plots the *mean isotropic spectrum levels* as a function of frequency for various sea states. Departures from the average of up to ±10 dB are quite common, particularly in shallow water and close to the surface. The ambient noise falls at about 5 dB per octave for frequencies above 500 Hz.

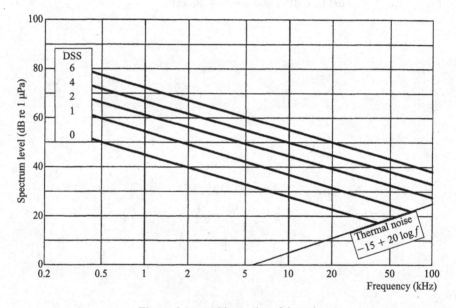

Figure 5.1   Ambient noise of the sea

Note the thermal noise plotted in Figure 5.1 where it forms a lower bound to the family of sea state noise curves. Sea state, wave height and wind speed are given in Table 5.1.

Table 5.1   Sea state, wave and wind speed

Sea state	0	1	2	3	4	5	6
Wind speed (knots)	≤1	5	13	16	19	22	28
Wave height (m), crest to trough	0	0.05	0.4	0.7	1.3	2	3

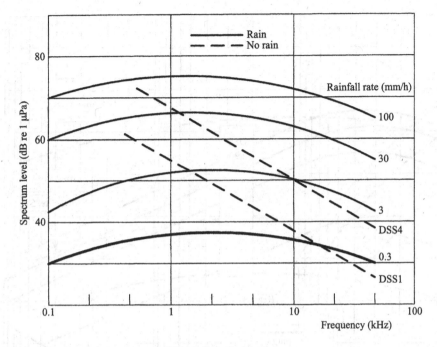

Figure 5.2   Rain noise

Rain can significantly increase the ambient noise. The spectra are quite flat with frequency, and Figure 5.2 plots theoretical values which agree quite well with measurements.

Shipping noise is important, particularly at frequencies below about 500 Hz. In harbours and harbour approaches, the dominant noise is from shipping – particularly small craft whose noise extends to frequencies of several kilohertz – and from industrial activity along the shore.

Biological noise is produced by a variety of marine life. Besides the individual sounds of sea mammals, snapping shrimps are a notable noise source. They produce quite a flat spectrum of noise between 500 Hz and 20 kHz, and it can be as high as 70 dB re 1 μPa.

Figure 5.3 plots spectrum levels for these noise sources (except biological). Some broad practical conclusions can be drawn:

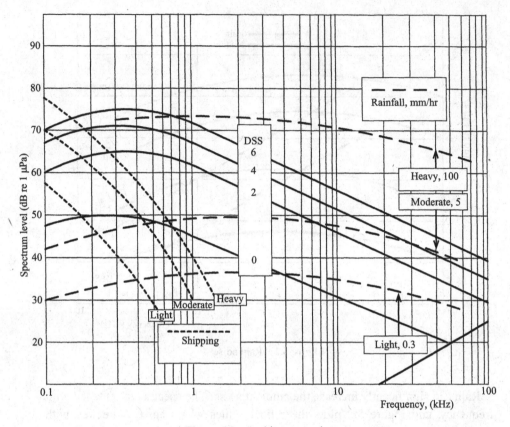

Figure 5.3   Ambient sea noise

- Heavy rain can dominate at all frequencies, but rainfall less than about 3 mm per hour is not significant.

- Noise resulting from the agitation of the sea due to wind and waves is dominant above 500 Hz. It is difficult to relate the actual noise level to the prevailing wind speed or sea state; the wind speed may be measured but the sea state is often based on a subjective judgement of wave height, and although wind speed and sea state are clearly related, rarely do their maximum or minimum values coincide in time.

- Shipping noise is unimportant above 1 kHz (radiated noise itself falls rapidly with frequency and also falls with range due to absorption losses). The noise below 1 kHz is likely to be dominated by discrete tones, but if there are many ships then the noise may appear to have a continuous spectrum.

5.4 Noise from a Vessel

There are three principal sources of noise from a vessel (which is taken to mean a surface ship, a submarine or a torpedo):

- Propulsion machinery and auxiliary machinery
- The propellers
- Flow noise

Figure 5.4 shows the relative contributions of these sources of self-noise at operational bearings for a typical surface ship at 10 kHz, particularly one where little has been done to reduce its radiated noise. Modern frigates can improve significantly on these figures, perhaps achieving SS2 at speeds up to 10 knots. The self-noise of a sonar is the sum of all noise sources, ambient and vessel, that are present at the array/water interface.

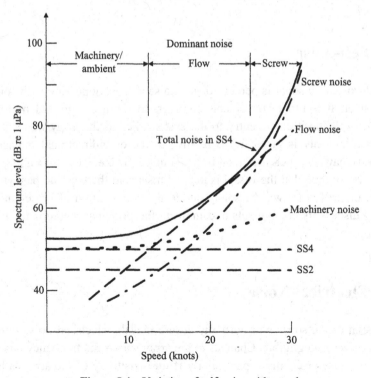

Figure 5.4 Variation of self-noise with speed

5.5   The Sonar Environment

Sonar arrays – mounted within domes or behind acoustic windows – find themselves in potentially quite noisy environments which can severely limit the performance of the sonar. These noise sources have been described above and fall within three classes:

- The internal noise of the ensemble of hydrophone and receiver

- The ambient noise of the sea

- Own vessel noise

The designer will ensure that the internal noise of the sonar is negligible, nothing can be done about the ambient noise of the sea, and therefore the design of the total system – vessel, dome and array – must attempt to minimize the, usually dominant, own vessel noise. A practical aim is to attempt to match the ambient noise of the sea corresponding to, say, SS2 for surface ships up to operating speeds of 15 knots and somewhat better for submarines.

5.6   Self-noise

The self-noise of a sonar is perceived as a noise voltage appearing at the output of the receiver. It is more convenient, however, for both design and performance analysis, to define it at the entry to the system, i.e., at the array/water interface. The noise intensity is given in dB relative to the omnidirectional intensity of a plane wave having a pressure equal to 1 µPa in a 1 Hz band, the *isotropic spectrum level*. If we assume that the vessel is in a calm sea and there are no parasitic noise sources, the self-noise will be solely due to the combination of vessel and sonar. We now discuss its components and indicate the principal methods for reducing them.

5.7   Electrical Noise

The design of the sonar will normally ensure that the electrical noise introduced by its receiver is negligible. Changes to the array and/or the frequency band of the sonar can cause problems – particularly if the sensitivity of the array is lower – and the noise factor of the receiver must be checked and improved if necessary.

Electrical interference on the cables between array and receivers is frequently a problem. Effective electromagnetic screening of the cables at low frequencies is very difficult. Fortunately, this problem can be eliminated or alleviated by modern techniques where the analogue voltages from the hydrophones are converted into digital form close to the array.

5.8   Machinery Noise

Machines closest to the array are the most troublesome; these may be propulsion machines or auxiliary machines. There are several ways to reduce their contribution to self-noise:

- Design the machines with quietness as a major parameter.

- Isolate machines, shafts and piping from the hull using anti-vibration mounts.

- Isolate the array and dome from the hull; provide damping between the array/dome and the hull; dampen the hull itself.

- Separate arrays and machines; site the arrays well forward of the main propulsion machinery; avoid placing any auxiliary machinery close to arrays.

5.9   Flow Noise

The principal cause of flow noise is flow over the sonar dome and over the hull close to the dome. Here are four ways to reduce it:

- Good hydrodynamic design of the dome
- Smooth interfaces between dome and hull
- Reduced surface roughness of dome and adjacent hull
- Coating the dome and the hull: the hull coating should continue significantly beyond the dome

5.10 Propeller Noise

Propeller noise is produced by a purely hydrodynamic mechanism such as cavitation at the tips of the blades or cavitation on the blades themselves, or by mechanical vibration of the blades. Here are two ways to reduce its contribution to self-noise and radiated noise:

- *Agouti*: a device which emits air bubbles in the vicinity of the blades to replace the water vapour bubbles created by cavitation. An improvement of the order of 10 dB is possible at high ship speeds (above 20 knots).

- *Baffles*: these are mounted within the dome to shield the array from propeller noise. They inevitably reduce the performance of the sonar over the stern arc protected by the baffle; a useful rule of thumb is to assume the allowable propagation loss is reduced by 10 dB over the stern arc.

5.11 Variation with Speed

Refer to Figure 5.4. At slow speeds (up to about 10 knots) *machinery noise* is dominant unless the vessel is particularly quiet or in high sea states, when the ambient noise of the sea will take precedence. *Flow noise* is dominant at medium speeds (10–20 knots), above which the *propeller noise* begins to dominate the self-noise even though flow noise continues to increase with speed. *Cavitation* at the dome can also be important at high speeds, particularly if the dome is poorly finished or has been damaged.

5.12 Variation with Frequency

The isotropic spectrum level of self-noise is observed to fall with frequency at about 6 dB per octave (proportional to $1/f^2$). At frequencies less than about 500 Hz this simple relationship should be used with care. The noise spectrum at low frequencies is unlikely to be continuous and is influenced by marine traffic and discrete frequency components.

5.13   Directivity

Self-noise is markedly directive. There is clearly a maximum over the stern arc, even with an intervening baffle. Measurements are frequently made in octave bands and averaged over operational bearings (the forward 270°), thus avoiding any bias which would result from including measurements astern.

5.14   Self-noise and Radiated Noise

We need to be clear on the distinction between self-noise and radiated noise:

- *Self-noise*: the total noise measured by a sonar system which forms the back-ground to detection of a wanted signal.

- *Radiated noise*: the noise radiated by a vessel and measured at some distance from the vessel, typically between 100 and 1000 m. It is the source of signals for passive sonars (Chapter 8).

Self-noise and radiated noise have many common sources and are frequently simply two different aspects of the same phenomenon. For example, propeller noise may dominate radiated noise at all speeds but only be an important contributor to self-noise at higher speeds; flow noise may dominate the self-noise when its source is close to the array but be an insignificant contributor to the radiated noise.

5.15   Addition of Noise Levels

The total noise level from two sources of noise is obtained by their incoherent addition on a power basis as given by the equation

$$P_T = 10 \log \left(10^{P_1/10} + 10^{P_2/10} \right)$$

where P_T is the total noise power and P_1 and P_2 are the individual noise sources, all in dB re 1 µPa.

- When $P_1 = P_2$ then $P_T = P_2 + 3$ dB
- When $P_1 = P_2 + 6$ dB then $P_T = P_1 + 1$ dB

Therefore if one noise level exceeds the other by at least 6 dB, the other makes a negligible contribution to P_T. The equation is plotted in Figure 5.5.

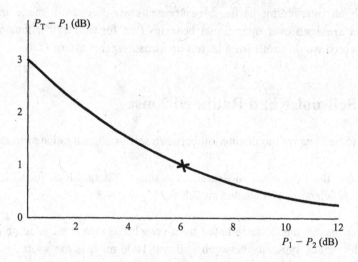

Figure 5.5   Increase in total noise level due to a weaker noise

5.16   Receiver Noise Factor

The receiver noise factor, NF_r, is defined as

$$NF_r = \frac{N_{out}}{N_{min}}$$

where N_{out} is the noise power at the output of the receiver and N_{min} is the unavoidable output noise, i.e., the noise at the input of the receiver due to the thermal noise of the generator connected to the receiver input multiplied by the receiver gain.

5.17 Noise Factor of a Sonar

Consider first the hydrophone placed in 'dead' water. The receiver is fed from the equivalent resistance of the hydrophone (Figure 5.6). The equivalent resistance of the hydrophone has two components, the *motional resistance*, R_m, defined in Section 5.2, and a *resistance R_p*, representing the losses in the hydrophone.

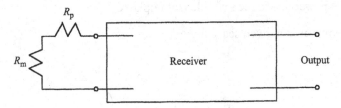

Figure 5.6 Hydrophone and receiver

The efficiency, η, of the hydrophone is given by

$$\eta = \frac{R_m}{R_m + R_p} = \frac{R_m}{R}$$

where $R = R_m + R_p$ is the total hydrophone resistance.

The noise factor of the *receiver alone* is NF_r. The noise factor of the *receiver plus hydrophone*, NF_{rh}, is greater. This is because, in defining NF_r, the thermal noise due to R is unavoidable, whereas for the ensemble (receiver plus hydrophone) the only unavoidable noise is due to R_m (which is smaller than R and would apply to a perfect hydrophone where $R_p = 0$).

The noise power, N_{min}, is therefore reduced by $\eta = R_m/R$, hence

$$NF_{rh} = \frac{NF_r}{\eta} \quad \text{or} \quad NF_{rh} = \frac{N_{rh} + N_{min}}{N_{min}}$$

Now consider the case where the sea state is not negligible. The noise factor becomes NF_s (at a given frequency), and

$$NF_s = \frac{\text{isotropic spectral noise intensity}}{\text{thermal noise (min)}} = \frac{N_s + N_{min}}{N_{min}}$$

and the resultant noise factor for the sonar, NF_{res}, for a given sea state, will be

$$NF_{res} = \frac{N_{output}}{N_{min}}$$

and since

N_{min} = unavoidable noise power

N_{rh} = noise power due to receiver plus hydrophone

N_s = noise power at a given sea state

we obtain

$$NF_{res} = \frac{N_{min} + N_{rh} + N_s}{N_{min}}$$

Using the expressions for NF_s and NF_{rh} obtained above, we can deduce the following expressions for the overall noise factor for the sonar in a given sea state, NF_{res}:

$$\boxed{NF_{res} = NF_s + NF_{rh} - 1 \quad \text{or} \quad NF_{res} = NF_s + (NF_r/\eta) - 1}$$

5.18   Acceptable Receiver Noise Level

The sonar designer is faced with the problem of determining an acceptable noise factor for the sonar receiver. If the receiver plus hydrophone were perfect, we would have $NF_{rh} = 1(0 \text{ dB})$ and therefore

$$NF_{res} = NF_s + NF_{rh} - 1 = NF_s$$

which is the minimum value of NF_{res} for a given sea state. We lose, therefore, due to the noise introduced by the sonar itself, in the ratio

$$\frac{NF_{res}}{NF_s} = 1 + \frac{1}{NF_s}\left(\frac{NF_r}{\eta} - 1\right)$$

Example 5.1
Consider a sonar operating at 5 kHz on a frigate whose self-noise is unlikely to fall below the ambient noise, which is 50 dB re 1 µPa for SS2.

We have

$$10 \log NF_s = 50 \text{ dB}$$

since $N_s \gg N_{min}$. Therefore

$$\frac{NF_{res}}{NF_s} = 1 + \frac{1}{100\,000}\left(\frac{NF_r}{\eta} - 1\right)$$

Suppose that we only lose 1 dB due to the internal noise of the sonar, then

$$\frac{NF_{res}}{NF_s} = 1.26 \quad \text{and} \quad \frac{NF_r}{\eta} = 0.26 \times 100\,000 = 44 \text{ dB}$$

If the efficiency of the hydrophone is 40 per cent (-4 dB) then the noise factor of the receiver itself will need to be less than 39 dB, which is easy to achieve.

The hydrophone sensitivity is given by $S_h = 20 \log v - 20 \log p$, and if $S_h = -200$ dB/V then

$$20 \log v = -200 + 20 \log p \quad \text{(dB)}$$

$$= -200 + 40 = -160 \text{ dB/V}$$

$$v = 10 \text{ nV}$$

i.e., the spectrum level of the receiver noise, referred to the input, must be less than 10 nV.

Example 5.2

Now consider a sonar, again operating at 5 kHz, installed on a submarine whose self-noise, in its quietest state, can be less than the ambient noise of the sea, and take the ambient noise in sea state 0 as 35 dB re 1 µPa.

We have

$$10 \log \text{NF}_s = 35 \text{ dB}$$

since $N_s \gg N_{\text{min}}$. Therefore

$$\frac{\text{NF}_{\text{res}}}{\text{NF}_s} = 1 + \frac{1}{3162} \left(\frac{\text{NF}_r}{\eta} - 1 \right)$$

Again suppose that we only lose 1 dB due to the internal noise of the sonar, then

$$\frac{\text{NF}_{\text{res}}}{\text{NF}_s} = 1.26 \quad \text{and} \quad \frac{\text{NF}_r}{\eta} = 0.26 \times 3162 = 29 \text{ dB}$$

and for the same hydrophone, the noise factor of the receiver has to be less than 25 dB:

$$20 \log v = -200 + 20 \log p \quad \text{(dB)}$$

$$= -200 + 25 = -175 \text{ dB/V}$$

$$v = 2 \text{ nV}$$

i.e., the spectrum level of the receiver noise, referred to the input, must be less than 2 nV.

5.19 Alternative Calculation

Section 5.18 gave a somewhat complex procedure for calculating the required spectrum noise level at the input to a sonar receiver. Here is a simpler approach that produces the same results.

Example 5.1 by simpler approach

Consider a sonar operating at 5 kHz on a frigate whose self-noise is unlikely to fall below the ambient noise, which is 50 dB re 1 µPa for SS2.

For 1 dB loss due to receiver noise, and again assuming 4 dB loss in the hydrophone, the spectrum level of the receiver noise must be

$$50 - 6 - 4 = 40 \text{ dB}$$

From Figure 5.5, 1 dB loss requires the receiver noise to be 6 dB down. Now we substitute 40 dB into the sensitivity equation:

$$20 \log v = -200 + 20 \log p \qquad \text{(dB)}$$
$$= -200 + 40 = -160 \text{ dB/V}$$
$$v = 10 \text{ nV}$$

i.e., the spectrum level of the receiver noise, referred to the input, must be less than 10 nV.

Example 5.2 by simpler approach

Now consider a sonar, again operating at 5 kHz, installed on a submarine whose self-noise, in its quietest state, can be less than the ambient noise of the sea, and take the ambient noise in sea state 0 as 35 dB re 1 µPa.

For 1 dB loss due to receiver noise, and again assuming 4 dB loss in the hydrophone, the spectrum level of the receiver noise must be

$$35 - 6 - 4 = 25 \text{ dB}$$

Therefore

$$20 \log v = -200 + 25 = -175 \text{ dB/V}$$

$$v = 2\text{nV}$$

i.e., the spectrum level of the receiver noise, referred to the input, must be less than 2 nV.

5.20   Practical Values

Table 5.2 gives practical values for performance comparisons and preliminary system design. All values are isotropic spectrum levels in dB re the intensity of sound due to a pressure of 1 μPa. The ambient noise figures apply to sonobuoys, helicopter dipping sonars, towed arrays and submarines. The self-noise figures apply to hull-mounted surface ship sonars; they are based on a ship moving at speeds up to about 15 knots in sea states of about 1 or 2 (low) and 4 or 5 (high).

Table 5.2   Practical noise values for design and performance comparison

	Isotropic spectrum level (dB) re the intensity of sound due to a pressure of 1 Pa							
	0.5 kHz	1 kHz	2 kHz	4 kHz	8 kHz	16 kHz	32 kHz	64 kHz
High ambient noise (SS4/5)	75	70	65	60	55	50	45	40
Low ambient noise (SS2)	65	60	55	50	45	40	35	30
High self-noise	84	78	72	66	60	54	48	42
Low self-noise	74	68	62	60	50	44	38	32

At frequencies below 500 Hz

- Ambient noise tends to flatten but can be increased by shipping.

- Self-noise is no longer a continuum, but is strongly influenced by discrete tones from propellers and machinery.

At frequencies above 64 kHz

- The thermal noise of the sea is increasingly the dominant source of noise and reverses the steady falls in both ambient and self-noise with frequency.

5.21 Problems

5.1 A hydrophone with a sensitivity of -170 dB/V receives over a 1000 Hz band centred on 100 kHz. What would its output be in SS2?

5.2 The self-noise of a sonar is 55 dB. If the ambient noise of the sea is also 55 dB, what would be the level in a third-octave band centred on 6 kHz measured at the output of a receiver beam of DI $= 20$ dB?

6
Reverberation

6.1 Sources of Reverberation

When sound is transmitted underwater it is scattered by marine life, inanimate matter distributed in the sea and the inhomogeneous structure of the sea itself, as well as by reflection from the surface and the sea bed. The component of the incident sound energy reflected back to the source is known as *backscattering*. This backscattered energy is *reverberation*, which comprises both the background to detection of a target and echoes from the target itself. Target echoes are simply a special case of reverberation.

6.2 Scattering and Reflection

Scattering and reflection occur wherever there is a change in Z, the specific acoustic impedance. The proportion of the energy reflected is given by

$$C_r = \frac{Z_r - Z_w}{Z_r + Z_w}$$

where Z_w is the specific acoustic impedance of water, Z_r is the specific acoustic impedance of the reflector and C_r is the coefficient of reflection at the boundary. When a reflection occurs at the *sea surface*, we have

- $Z_r = 415 \, \mathrm{kg\,m^{-2}\,s^{-1}}$
- $Z_w = 1.5 \times 10^6 \, \mathrm{kg\,m^{-2}\,s^{-1}}$
- $C_r = -0.9995$

Here Z_r is the specific acoustic impedance of air and Z_w is the specific acoustic impedance of water. Almost all of the sound is reflected and only a tiny fraction, 0.0005, escapes into the atmosphere. Note the phase change.

When a reflection occurs at the *sea bed*, the situation is much more complicated. Because Z_r, the specific acoustic impedance of the sea bed, is now *much greater* than Z_w, the specific acoustic impedance of water, the value of C_r is positive (but still almost unity) and there is no phase change at the reflection. Once again, much of the sound is reflected and only a fraction penetrates the sea bed.

The sea bed, however, has highly variable acoustic properties due to its composition and inhomogeneous nature, with the result that Z_r, the specific acoustic impedance of the sea bed, changes markedly with both depth of penetration and laterally. A sufficiently accurate and detailed knowledge of the bottom to ensure useful predictions for scattering, and for propagation within the bottom itself, is never realistically available.

6.3   Boundary Roughness

The nature of the reflection and scattering depends on the degree of roughness of the boundary. If the incident sound wave strikes a perfectly smooth sea surface or sea bed, nearly all the energy is reflected in the *specular direction*. As the boundary roughens, more and more sound energy is scattered in non-specular directions and the specular reflection is reduced in intensity until, for a very rough surface, there is no discernible peak in the specular direction at all.

The roughness of a boundary is indicated by the Rayleigh parameter, R_p, given by

$$R_p = kh \sin \theta$$

where h is the wave height, measured from trough to crest, θ is the *grazing angle* and $k = 2\pi/\lambda$ is the wave number. In practice when $R_p \ll 1$ the surface is taken to be smooth and when $R_p \gg 1$ the surface is rough.

Figure 6.1 illustrates the reflection and scattering of sound from (a) perfectly smooth boundaries and (b) rough boundaries. The boundary shown is the sea surface; inversion of the figure represents the similar situation when the boundary is the sea bed.

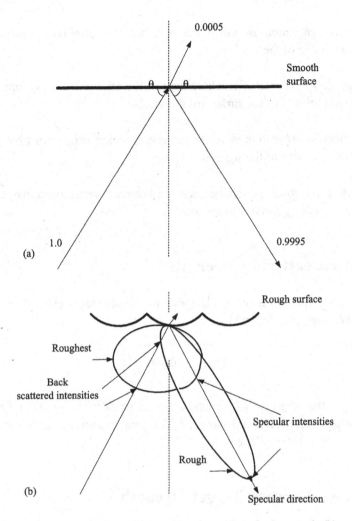

Figure 6.1 Scattering of sound from a boundary: (a) perfectly smooth, (b) rough

6.4 Classes of Reverberation

Along with noise, reverberation – the backscattered part of the total reflection – forms the background to the detection of targets by an active sonar. In many environments, reverberation can be the dominant component of this background. It is therefore very important for the design of active sonars to have a good measure of its magnitude and how it varies with roughness and grazing angle.

Reverberation may be classified according to its source:

- *Volume reverberation* is where the reverberation originates from scatterers within a volume of the sea.

- *Sea surface reverberation* is where the reverberation originates from scatterers spread over an area of the surface of the sea.

- *Sea bottom reverberation* is where the reverberation originates from scatterers spread over an area of the sea bed.

For analytical purposes, sea surface and sea bottom reverberation may be lumped together as *boundary (area) reverberation*.

6.5   Backscattering Strength

Backscattering strength is the fundamental parameter that decides the intensity of the reverberation. It is defined by

$$S_{s,v} = 10 \log \frac{I_{scat}}{I_i}$$

where I_{scat} is the intensity of the sound scattered (back to the source) by a unit area or unit volume, referred to a distance of 1 m from its acoustic centre, and I_i is the intensity of the incident plane wave.

6.6   Reverberation Target Strength

The target strength for reverberation, TS_R, is a measure of the background to detection of a target and is used later to derive the reverberation-limited active sonar equation.

$$TS_R = S_{s,v} + 10 \log A, V$$

where A and V are the total reverberating area or volume, respectively, as defined by the 'two-way' beamwidth of the sonar – the effective beamwidths of the projector plus hydrophone combination in both azimuth and elevation. For example, the horizontal two-way beamwidth of the combination of an omnidirectional, in azimuth, projector and a 10° , in azimuth, receive array is 10° (the smaller of the two).

6.7 Volume Reverberation

To determine TS_R for volume reverberation, it is necessary to determine $10 \log V$. The total reverberating volume (Figure 6.2) is given by

$$V = \int_V B_r B_t \, dV$$

where B_r and B_t are the receive and transmit beam patterns.

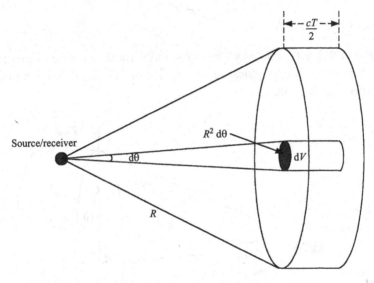

Figure 6.2 Reverberating volume

The length of the cylinder, dV (its extent in range), is such that the reflections from all scatterers within it return to the source simultaneously. (The scattering of the front end of the pulse by the rear scatterers in dV returns to the source at the same time as the scattering of the rear end of the pulse by the front scatterers in dV.)

The extent in range is therefore $cT/2$, where T is the pulse length and c is the speed of sound. (Note the division by 2, a constant source of confusion between active, two-way, and passive, one-way, sonars.) The volume can now be written

$$V = \frac{cT}{2} R^2 \int_V B_r B_t \, d\theta$$

The integral is the equivalent beamwidth of the two-way (transmit/receive) combination. In this expression T is either

- The actual pulse length, for a CW pulse
- The reciprocal of the bandwidth, for a wideband pulse (see later)

and if R is large compared with the cross-section of the volume (say $R > 4R\theta$), the volume is given by

$$V = \frac{cT}{2}\frac{\pi R^2 \theta_h \theta_v}{4}$$

where θ_h and θ'_v are the two-way horizontal and vertical beamwidths expressed *in radians*. Figure 6.3 is a graphical representation of this simplified expression for the reverberating volume.

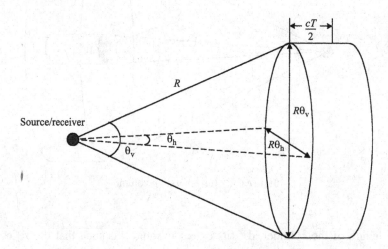

Figure 6.3 Reverberating volume

Example 6.1
A sonar has two-way horizontal and vertical beamwidths of 12° (0.2 rad) and transmits a pulse of length 1 ms. What is the TS_R at 1000 m for $S_v = -80$ dB?

We calculate

$$V = \frac{1500 \times 10^{-3}}{2} \times \frac{\pi \times 10^6 \times 0.2^2}{4} = 2 \times 10^4$$

and

$$\text{TS}_\text{R} = -80 + 10\log 2 \times 10^4 = -80 + 43 = -37 \text{ dB}$$

When the beam patterns intercept a boundary, i.e., the surface or the bottom of the sea – which, except in deep water, occurs at quite close ranges (perhaps 1000 m) – the volume will be smaller than given by the formula. It is then possible, but not usually necessary, to work out the cross-sectional area from a knowledge of the depth of water and the depth of the source and to use this to determine the volume.

However, as soon as a boundary is intercepted by both transmit and receive beam patterns, the boundary (area) reverberation is almost invariably dominant and volume reverberation is no longer of interest.

6.8   Boundary Reverberation

To determine TS_R for boundary reverberation we need to determine $10 \log A$, where A is the area of the scattering boundary (Figure 6.4). Provided the axis of the beam is only slightly inclined towards the scattering boundary, which is usually the case, it is only necessary to consider the horizontal beam patterns of the arrays, the solid angle for the volume case reduces to the *horizontal* beamwidth, and A becomes

$$A = \frac{cT}{2} R\theta_h$$

with θ_h in *radians*.

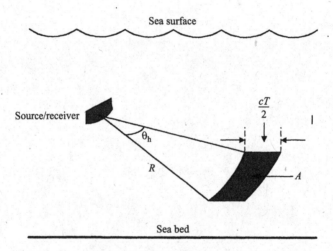

Figure 6.4   Boundary reverberation

Example 6.2
For the same sonar, what is the TS_R at 1000 m for $S_s = -30$ dB?

We calculate

$$A = \frac{1500 \times 10^{-3}}{2} \times 10^3 \times 0.2 = 150 \text{ m}^2$$

and

$$\text{TS}_R = -30 + 10 \log 150 = -30 + 22 = -8 \text{ dB}$$

- In deep water the dominant reverberation will be from the sea surface.

- In shallow water (< 200 m) the dominant reverberation will be from the sea bed unless the wind speed is high (see later).

- Volume reverberation is seldom dominant; the exception is in calm, deep seas at long range.

6.9 Scattering Layers

Some scatterers in the sea – such as the deep scattering layer, or a layer of bubbles below the sea surface – lie in layers of limited thickness and are best considered as a modified form of *boundary* reverberation. Quite wrong results would be obtained if the scattering from these sources were considered to be uniformly present throughout the entire volume, as previously calculated.

If the volume scattering strength of the *layer* is S_v and the thickness of the layer is H metres, then the scattering strength of the layer, S_ℓ, is given by

$$\boxed{S_\ell = S_v + 10 \log H}$$

and this value must be used in the expression for *boundary* reverberation.

6.10   Volume Scattering Strength

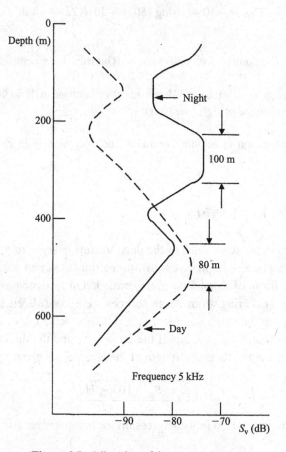

Figure 6.5   Migration of deep scattering layer

Volume scatterers are not uniformly distributed in the sea and tend to be concentrated in the deep scattering layer (DSL). This source of reverberation is overwhelmingly biological in nature and is a complex mix of different organisms resulting in scattering strength versus depth profiles that change with frequency, location and time. The representative S_v profiles in Figure 6.5 show vertical migration of the DSL with time of day. Estimates of thickness are shown and

$$S_\ell \, (\text{day}) = -76 + 10 \log 80 = -57 \text{ dB}$$

$$S_\ell \, (\text{night}) = -74 + 10 \log 100 = -54 \text{ dB}$$

6.11   Sea Surface Scattering Strength

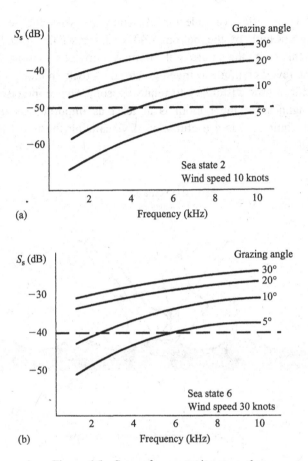

Figure 6.6   Sea surface scattering strengths

The roughness of the sea surface, and the possibility of bubbles trapped just beneath it, results in significant scattering of sound. The scattering strength varies with the grazing angle of the incident wave and with frequency. Empirical formulae by Chapman and Harris fit the many measurements quite well, particularly at frequencies between 1 and 10 kHz. Figure 6.6 plots these formulae for two representative sea states. Note that S_s increases markedly with grazing angle and also with frequency (around 3 dB per octave at the lower grazing angles which are of particular interest to the sonar designer). The dashed lines show recommended values for use in sonar design, representing high sea states (-40 dB) and low sea states (-50 dB).

6.12   Bottom Scattering Strength

The sea bed is an effective reflector of sound. As with the sea surface, the backscattering strength of the bottom (S_b) varies with the grazing angle and frequency and also with the nature of the bottom. Early observations indicated that the level of the reverberation was much greater over rocky bottoms than over sand and mud, and it is now customary to relate S_b, somewhat imprecisely, to bottom type. The roughness of the bottom is at least as important as its constituent materials and Figure 6.7 is a useful way of visualizing the redistribution of the incident sound.

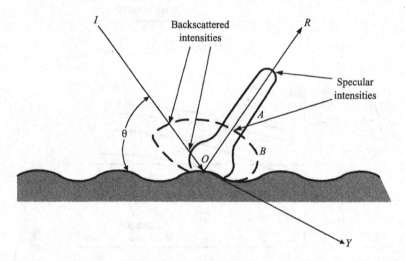

Figure 6.7   Sound scattering patterns

Figure 6.7 shows two directivity patterns for the sound incident upon the bottom at point *O*. The radius of the patterns in any direction is proportional to the intensity of the sound reflected in that direction.

- *Pattern A* is characteristic of a *smooth* bottom, where the specular reflection in direction *OR* is large and the scattering in other directions including the backscattering back to the source in the direction *OI*, at a grazing angle, θ, is small.

- *Pattern B* is characteristic of a *rough* bottom, where the specular reflection in direction *OI* is much greater. (The backscattering strength, S_b, is much greater.) Sound is also absorbed where it penetrates the bottom in direction *OY*.

Many, frequently discordant, measurements have been reported in the literature and this, coupled with a lack of detailed knowledge of the sea bed at any particular location, results in considerable uncertainty over the value of S_b to use for a particular problem.

Bottom scattering strengths may be considered roughly constant at frequencies up to 10 kHz and grazing angles up to 10°. They are, however, very dependent on the material of the sea bed, varying from a low value of −45 dB for mud to −25 dB for rock, as shown by the results from a survey of UK shallow water areas (Figure 6.8).

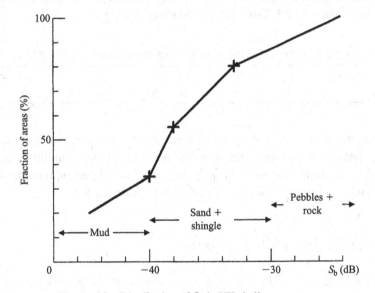

Figure 6.8 Distribution of S_b in UK shallow water areas

Bottom scattering strengths are generally greater than those from the surface and therefore, when the geometry is such that the sonar vertical beam patterns intercept the bottom (which is always the case for shallow water at practical ranges), then bottom reverberation is often the dominant background to detection and classification.

Figure 6.8 illustrates the wide range of S_b which may be encountered in shallow waters. We shall see later that S_b is an extremely important factor in determining the performance and design of an active sonar and if, during its design, an incorrect value is chosen for the maximum S_b likely to be encountered, the sonar's performance will not be optimum.

Over a uniform bottom the basic detection performance will be determined by the average value of S_b, which is little affected by comparatively rare, submarine-like false alarms. The bottom may be mud, clay, sand, shingle, pebbles or a mix of these and the calculated S_b (derived from measurements of reverberation levels) falls within the range -35 to -50 dB at the low grazing angles encountered in shallow water.

Over a non-uniform bottom (rocky outcrops, shorelines, undulations) the average value of S_b is greatly influenced by such local features and it is suggested that the actual values of S_b, used to determine the initial detection ranges of a sonar, should be some 5–10 dB *less* than the average measured values. Thus, values for S_b which are greater than -35 dB, i.e., falling within the range -35 to -25 dB, and typical of non-uniform sea beds, should be reduced by 5 dB but never to less than -35 dB. Table 6.1 should make this clear.

Table 6.1   Reduce S_b values by 5 dB but never to less than -35 dB

S_b measured (dB)	-25	-27	-29	-31	-33	-35
Use (dB)	-30	-32	-34	-35	-35	-35

This argument is supported by initial detection trials where, over such difficult bottoms, final detection and classification are delayed by a multitude of false alarms but post-trials analysis frequently reveals missed detection opportunities at greater ranges (consistent with postulating lower values for S_b).

6.13   Variation with Frequency

At frequencies above 10 kHz, S_b appears to increase at about 3 dB per octave for the smoother (mud, sand) bottoms, i.e., where the roughness is small compared to a wavelength; but for rougher (shingle, pebbles, rock) bottoms, S_b is independent of frequency. This frequency dependency is an unfortunate result for mine hunting sonars which, to achieve sufficient definition, must operate at high frequencies.

6.14 Reverberation under Ice

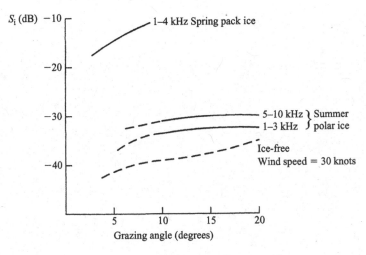

Figure 6.9 Under ice scattering strengths

The underside of the ice cover in polar regions has a very high backscattering strength, S_i, when it is rough and non-uniform (e.g., spring pack ice). When the underside is smoother (e.g., summer polar ice), S_i is significantly less. Most of the S_i measurements show an approximately 3 dB per octave frequency dependency and are generally greater than S_s measurements in ice-free water (Figure 6.9).

6.15 Problem

6.1 A sonar system has a horizontal beamwidth of 10° and transmits a pulse of duration 100 ms. If the backscattering strength is −40 dB, what is the reverberation target strength, TS_R, at 4000 m? What is the TS_R at twice the range?

7

The Sonar Equations

7.1 What Are They?

The sonar equations relate the parameters of the entire sonar system – the sonar and its platform, the target and the environment – in order to determine its performance. This chapter will develop the basic equations and explore the concepts of detection threshold (DT), detection index, d, and their relationships to the probability of detection, P_d, and the probability of false alarms, P_{fa}. Later chapters will develop the sonar equations to make them applicable to active, passive, intercept and communications sonars. This is because it requires a detailed knowledge of the signal processing to develop an expression for DT appropriate to a particular class of sonar.

7.2 What Are Their Uses?

The sonar equations are used to predict the performance of a known design, and to design a sonar for a given performance. For performance prediction, the parameters of the sonar are known and an estimate of its detection performance is required. The sonar equations are solved for the allowable propagation losses at specified probabilities of detection and false alarms, and the allowable propagation losses are then converted to estimates of detection range in the environments of interest.

When designing a sonar to detect at some predetermined range, the propagation loss is first estimated for the environment of interest and is then used in the sonar equations to solve for those parameters over which the sonar designer has some control. This process is often severely constrained by, for example, platform size or equipment cost and is considerably aided by the intuition of an experienced sonar designer.

7.3   The Basic Sonar Equation

The basic sonar equation simply expresses the difference between signal-to-noise ratio at the output of the beamformer and the *detection threshold* (DT). This difference is the *signal excess* (SE) and in dB form it is given by

$$SE = S - N - DT$$

Signal power is in the analysis bandwidth and noise power is in a 1 Hz band. This threshold is defined such that a signal with $S - N$ (at the output of the beamformer) equal to DT, has a specified (often 50 per cent) probability of detection (P_d) for a required probability of false alarms (P_{fa}). The value of DT depends on these probabilities and the sonar signal processing. Put another way, a signal excess of zero corresponds to a P_d of 50 per cent and a positive SE indicates $P_d > 50$ per cent.

7.4   The Basic Passive Equation

Passive sonars detect signals radiated by the target. The signal level is therefore $S = SL - PL$, where SL is the source level of the target (the radiated noise in the frequency band of interest). The basic passive sonar equation is therefore

$$\boxed{SE = (SL - PL) - N = DT}$$

7.5   The Basic Active Equation

Active sonars detect target echoes. The signal level is therefore $S = SL + TS - 2PL$, where SL is now the source level of the projector array. Note the 'active' 2. The basic active sonar equation is therefore

$$\boxed{SE = (SL + TS - 2PL) - N - DT}$$

7.6 Detection Threshold and Detection Index

All sonar receiving systems conform to Figure 7.1. The *minimum discernible signal* (MDS) is the S/N ratio at the array face which results in preset P_d and P_{fa} values:

$$\text{MDS} = \text{DT} - \text{DI} \quad \text{(if the noise is isotropic)}$$

$$\text{MDS} = \text{DT} - \text{AG} \quad \text{(if the noise is directional)}$$

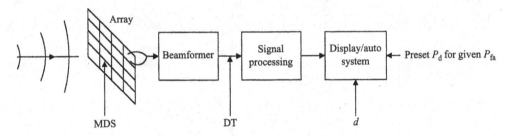

Figure 7.1 Basic sonar receiver

The *detection threshold* (DT) is the ratio of mean signal power to mean noise power (expressed in dB form) measured *after beamforming*, which results in the preset P_d and P_{fa} values. The signal power is in the analysis bandwidth and the noise power is in a 1 Hz band. *Detection index, d,* is defined as

$$d = \left(\frac{\text{mean } (S + N) - \text{mean } (N)}{\text{std dev}(N)} \right)^2$$

where the quantities are measured at the display or automatic system.

Figure 7.2 shows curves of probability density plotted for N alone and for $S + N$.

The *detection index*, d, is equivalent to the ratio $(S + N)/N$ of the envelope of the signal processing output where the threshold, T, is established. The area under the curve of $S + N$ above T is the probability that an amplitude in excess of T is due to signal plus noise; this area equals P_d. The area under the curve of N above T is the probability that an amplitude in excess of T is due to noise only; this area equals P_{fa}. These two probabilities therefore vary with the threshold setting and depend on the value of d.

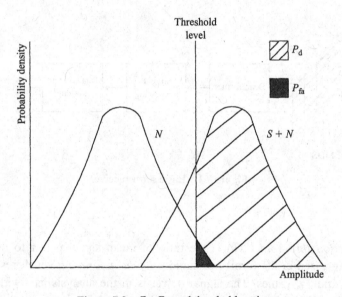

Figure 7.2 P_d, P_{fa} and threshold setting

7.7   Receiver Operating Characteristics

Curves of receiver operating characteristics, ROC curves, express P_d and P_{fa} as a function of $5 \log d$ for the appropriate distributions of N and $S + N$. The standard ROC curve (Figure 7.3) assumes Gaussian statistics for both distributions. Two

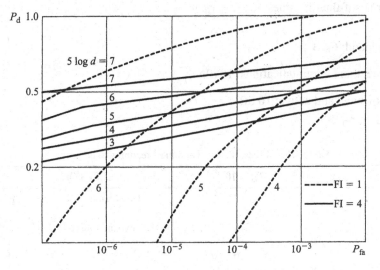

Figure 7.3   ROC curves for Gaussian statistics: N and $S + N$

levels of signal variability are shown, defined in terms of the *fluctuation index* (FI) given by

$$FI = \frac{\sigma(S + N)}{\sigma N}$$

FI $= 1$ corresponds to a stable signal where the variability in $S + N$ is entirely due to N. Increasing FI corresponds to increasing the signal variability.

7.8   ROC Curves

There is no one ROC curve applicable to all sonars. Figure 7.3, which assumes
Gaussian statistics, is applicable to sonars with large BT products (broadband
sonars). Table 7.1 gives values of $5 \log d$ taken from the *Sonar Modelling
Handbook* ROC curves together with their recommended usage. But here are some
useful rule-of-thumb values:

- Broadband: 6 dB

- Intercept and communications: 10 dB

- Narrowband and active: 10 dB

Table 7.1   Values of $5 \log d$ and their recommended usage

	$P_{fa} = 10^{-5}$	$P_{fa} = 10^{-6}$	USE FOR	
$P_d = 0.5$	6	7	Broadband	FI = 1
	6	7		FI = 4
	12	13	Narrowband power	IF = 1
	10	11		IF = 2
	9	10		IF = 4
	8	9		IF = 8
	8	9	Narrowband amplitude	
	10	11	Active, intercept and communications	
$P_d = 0.9$	8	8	Broadband	FI = 1
	$\gg 8$	$\gg 8$		FI = 4
	20	21	Narrowband power	IF = 1
	15	16		IF = 2
	13	14		IF = 4
	9	11		IF = 8
	12	13	Narrowband amplitude	
	12	13	Active, intercept and communications	

Integration Factor (IF) is defined in Section 8.5.

7.9   Problem

7.1 The P_d of a sonar is to be 0.5. What will be the change in P_{fa} which results from
reducing $5 \log d$ from 6 dB to 5 dB? Use Figure 7.3.

8

Passive Sonar

8.1 Radiated Noise

Radiated noise is the noise emitted by a vessel and received by a hydrophone, or an array of hydrophones, at some distance from the vessel. Radiated noise is the source of signals for *passive sonars* which are designed to detect radiated noise against a background of ambient and self-noise.

8.2 Source Level

The source level of radiated noise is analogous to the transmit source level for an active sonar and is used in the passive sonar equations. Measurements are made at some distance from the source (between 100 and 1000 m) and reduced to a standard distance of 1 m by assuming spherical spreading and, where significant, absorption. Measurements are normally made in third-octave bands over the frequency range of interest – from 10 Hz to 100 kHz – and converted to spectrum levels on the assumption that the intensity is constant over the band.

To investigate the *line structure* of the radiated noise spectrum, a much narrower bandwidth is required. Discrete lines are important up to about 3000 Hz and narrowband analysis is carried out over a selection of frequency bands of interest such as those in Table 8.1.

Table 8.1 Analysis bandwidths are narrower

Frequency band (Hz)	Analysis bandwidth (Hz)
0–100	0.3
100–1000	1
1000–3000	4

8.3   Nature of Radiated Noise

The noise radiated by a vessel exists as a continuous spectrum, or *continuum*, on which are superimposed *narrowband discrete* components, known simply as *lines* or *tonals*. Both the continuum and discrete components reduce in intensity as the frequency increases. Machinery noise and propeller noise dominate the spectra of radiated noise in most conditions. The lower frequency end of the spectrum is dominated by machinery lines and blade rate lines of the propeller. These lines die away with increasing frequency and become submerged in the continuous spectrum of propeller and machinery noise. As the speed of a vessel is increased, the continuum increases in intensity and extends to lower frequencies.

Figure 8.1 illustrates the effect of increasing speed. For submarines, a decrease in depth − because it increases any noise due to cavitation − has the same general effect. For a given speed (and depth for submarines and torpedoes), a broad *crossover frequency* exists, below which the spectrum is dominated by tonals and above which the spectrum is the continuum of the cavitating propeller.

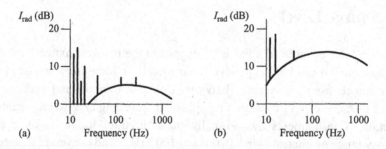

Figure 8.1   Radiated noise: continuum and tonals for (a) low speed and (b) high speed

For surface ships and submarines this frequency is between 100 and 500 Hz, depending on speed and, for submarines, depth. For torpedoes the crossover frequency is higher, perhaps between 500 and 1500 Hz, and the tonals extend to higher frequencies, say 3000 Hz, because of the higher speeds of machinery and propellers.

8.4   Practical Values

Information on the radiated noise of vessels, particularly submarines and torpedoes, is highly classified and the sonar designer must resort to the classified literature for details of levels and frequencies of tonals radiated from modern vessels.

Figure 8.2 shows representative radiated noise spectra for sonar targets. The radiated noise of a frigate is also shown for comparison. The spectra should be considered as 'worst case', particularly in the regions dominated by discrete lines.

Submarine and torpedo quietening measures have succeeded in significantly reducing or eliminating many discrete lines, particularly at the higher frequencies, resulting in the need for passive sonars to operate at ever lower frequencies

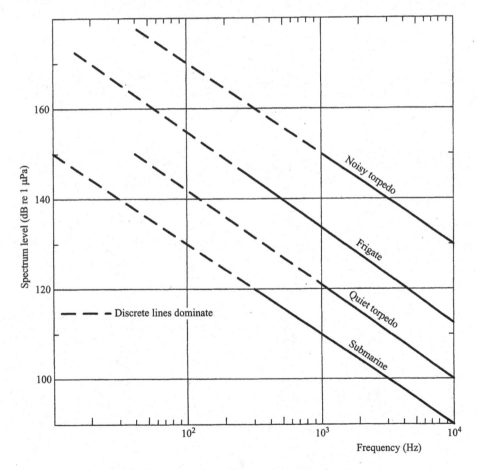

Figure 8.2   Representative radiated noise spectra for sonar targets

(therefore they need ever longer towed arrays) to detect the remaining lower frequency lines, which are difficult to eliminate.

An older submarine, for example, may have a tonal at 200 Hz of intensity 125 dB, but tonals near this frequency for a modern submarine will have been reduced to a value below the broadband continuum which tends to level off in intensity at, say, 300 Hz (and perhaps 1000 Hz for torpedoes).

8.5 Broadband and Narrowband

Passive sonar systems investigate the total spectrum of the noise radiated from targets, using broadband and narrowband techniques. Broadband sonars examine the total energy over a wide frequency band, which is generally divided into octaves. The *broadband* sonar equations, derived later in this chapter, indicate that performance is improved by *increasing* the bandwidth. This is only true if the bandwidth does not significantly exceed the spectrum of the noise radiated by the threat.

Narrowband sonars split the total energy into narrow frequency analysis cells in order to search for discrete radiated lines. The *narrowband* passive sonar equations, also derived later in this chapter, indicate that performance is improved by *reducing* the analysis bandwidth (until the signal is over resolved).

The broadband noise from a propeller may be amplitude modulated at the blade rate fundamental and harmonic frequencies. A technique known as DEMON (demodulation) exploits this by narrowband analysis over the band which covers these modulation frequencies.

Because they provide a detailed knowledge of the threat's radiated noise, narrowband sonars give good detection and classification capabilities. Because prominent tonals are radiated at comparatively low frequencies, particularly by submarines, bearing accuracy is limited when compared with broadband sonars operating at higher frequencies where beams are narrower and correlation techniques may be employed to further improve bearing accuracy. (Very long towed line arrays, however, achieve similar bearing accuracies at these low narrowband frequencies.)

Figure 8.3 shows a complete passive sonar system. The array signals are beamformed over the entire frequency range and output to audio, broadband and narrowband sonars. The *audio sonar* will have facilities to select a beam and a listening bandwidth.

The *broadband sonar* will detect signals from the beamformer (using energy or cross-correlation techniques); integrate (incoherently sum the energy in the signal band); normalize the signals and display them, typically in a bearing (beam) versus time format. The integration gain for energy detection is $5 \log BT_e$, and for cross-correlation $5 \log 2BT_e$, where B is the signal bandwidth (usually an octave) and T_e is the integration time (the time a signal is likely to remain within a beam).

The *narrowband sonar* performs a frequency spectrum calculation to convert the time series beamformer output into power spectral data in two stages. The first stage is applied to all data from all beams to produce a surveillance capability; the second stage allows the operator to 'zoom in' (increase the frequency resolution) to a part of the total spectrum. This second stage is also known as vernier processing. The gain from the total process is $10 \log B$, where B is the final analysis bandwidth. If

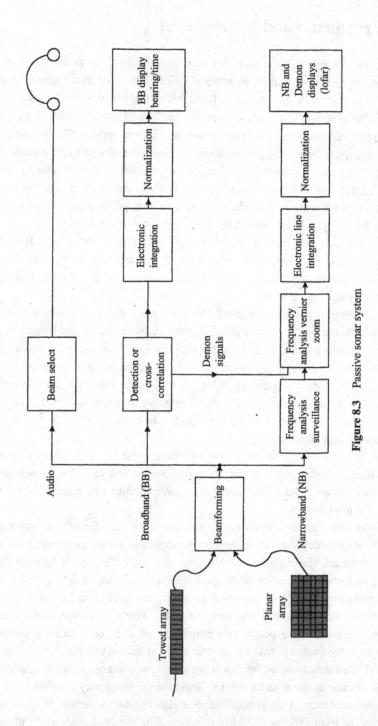

Figure 8.3   Passive sonar system

the signal bandwidth, B_s, is greater than B, the signal is over resolved and there will be a reduction in gain of $10 \log B/B_s$. Electronic integration sums the energy from each analysis cell over a number of time steps. The number of steps is known as the *integration factor* (IF) and the gain is $5 \log$ IF.

8.6 Normalization

There are significant differences in the mean noise levels with frequency, bearing (beam) and time, which must be reduced before display to the operator. Normalization schemes all depend on using the information surrounding the cell to be displayed to reduce these differences to the dynamic range of the display. Figure 8.4 shows an algorithm for normalizing a line of data in the frequency dimension before display. The same process is repeated for all analysis cells.

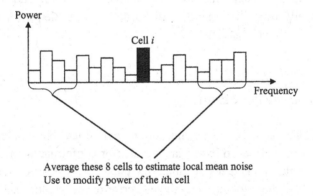

Figure 8.4   A typical normalization algorithm

8.7 A Note on Swaths

Swaths are signals with bandwidths significantly greater than the usual resolution of a narrowband processor, but much smaller than the octave or more bandwidths of broadband processors. Swath detection, therefore, requires wider analysis bandwidths and modifications to the trend removal or normalization algorithms by making similar increases in window widths and spacings.

8.8   Passive Arrays

Noise reduction programmes have succeeded in reducing the broadband noise continuum and reducing or eliminating the higher frequency tonals radiated by older designs of submarines and torpedoes. Modern passive sonars, therefore, must employ large arrays to achieve sufficient gain and directivity at the low frequencies necessary for the timely detection and classification of submarines and torpedoes.

Surface ship hull-mounted arrays

Surface ship hull-mounted arrays will generally take the form of cylindrical arrays mounted in bow or keel domes. Operating frequencies may extend down to about 100 Hz for narrowband tonals and up to about 10 kHz for broadband. The DI will clearly be small at frequencies below about 1 kHz and the self-noise high. Performance against submarines will be ineffective but torpedoes (apart from quiet, electrically propelled, marks) will be detected at adequate ranges for the deployment of countermeasures.

Submarine hull-mounted arrays

Submarine hull-mounted arrays again may be cylindrical but a submarine offers much better opportunities to install large planar or conformal arrays. The DI will still be small at the frequencies of submarine tonals, but the self-noise is generally significantly less than for surface ships. Performance against submarines and torpedoes is therefore effective against all but the quietest of submarines.

Line arrays

Line arrays towed from submarines or surface ships – by virtue of their length, typically 32 or 64 wavelengths, and low operating frequencies, down to about 10 Hz – are very effective against both submarines and torpedoes. Separation from the tow vessel assists in reducing the self-noise seen by the towed array, particularly at bearings other than those of the ahead beams which receive the stern radiated noise of the tow vessel in their main lobes.

8.9 Passive Aural

The principal characteristics of the human ear are as follows:

- Dynamic range 120 dB (see Section 1.12)

- Wide frequency range: 20 to 15 000 Hz

- A powerful frequency analyser

As a frequency analyser, the human ear can resolve two sounds which differ by as little as 50–100 Hz over much of the audible band. Because of these characteristics, the ear is still important in passive sonar, and perhaps unrivalled as the final arbiter in the classification process.

The ear is not a simple broadband listening channel, but rather may be likened to a contiguous comb of narrowband filters covering the entire audio spectrum. The bandwidth of each of these hypothetical filters is known as the *critical bandwidth* (Δf_c) of the ear and is between 50 and 100 Hz for frequencies between 300 and 2000 Hz and increasing somewhat outside this band.

The quantity Δf_c is the bandwidth such that an increase in spectrum level (within reasonable limits), *outside* this band has no masking effect on a signal *within* the critical band (Figure 8.5). Therefore, the masking noise is given by

$$C(f) = N_s(f) + 10 \log \Delta f_c(f)$$

where $N_s(f)$ is the spectrum level of the noise at signal frequency f.

Experimental verification has shown that the recognition level of the ear is close

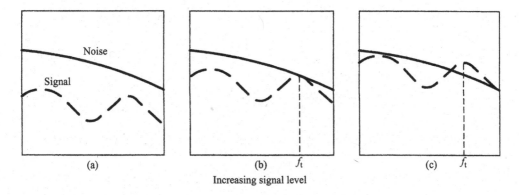

Figure 8.5   Masking of a signal (wanted noise) by noise

to $C(f)$ and, from the concept of critical bandwidth, it follows that as the level of a signal (which is in fact the wideband radiated noise emitted by the threat) is increased, as soon as this level reaches the noise level at any frequency within the receiver bandwidth, it is perceived (at $P_d = 50$ per cent) as a signal at that frequency, known as the *threshold frequency*, f_t.

Reducing the listening bandwidth risks eliminating the threshold frequency (or moving it to another value where the noise is higher) and therefore the audio bandwidth should cover the entire frequency spectrum of the expected signal.

The ear will have detected the signal at the level represented by Figure 8.5(b), but a simple broadband receiver will not detect the signal until the level is increased to about that represented by Figure 8.5(c), a positive signal-to-noise ratio.

8.10 Passive Displays

The ear has the advantage of being able to simultaneously use all the information contained within the signal – harmonics, waveform, fluctuations in intensity and frequency – which can often result in very low recognition differentials as well as aiding in threat classification. Visual indicators, however, also benefit from the quasi-permanent character of the signal by performing electronic integration after detection and visual integration at the display.

The *electronic integration time* is limited by the time a threat may be expected to stay within a beam – perhaps 2 s for a torpedo and 20 s for a submarine.

The *visual integration time* is not similarly limited because the integration can now take place through many beams, although clearly it cannot be longer than a few minutes for a torpedo and still offer a timely countermeasure response. The visual integration time is still limited by practical considerations such as the display capacity. There are also diminishing returns in the gain achieved; the theoretical 1.5 dB for each doubling of displayed lines tends to level off at about 10 dB for 128 lines.

Any of the traces in Figure 8.6 could be a subsurface contact. A link with radar (for surface ships) can eliminate surface contacts. Any remaining contact is highly likely to be a torpedo or a submarine. Submarine detection is unlikely using a surface ship hull-mounted array but possible using a submarine hull-mounted array (because the self-noise is lower and much larger arrays are practical). Long towed arrays deployed from either surface ships or submarines perform well against both torpedoes and submarines.

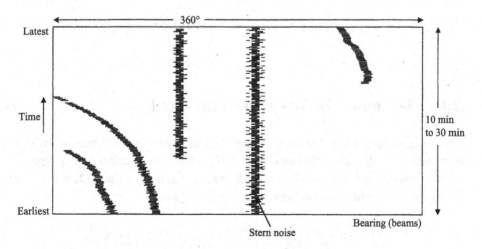

Figure 8.6 Broadband (bearing/time) display

Narrowband signals are displayed in a format known, for historical reasons, as a LOFARGRAM (low frequency analysis recording gram) where the outputs from all beams or a selected group of beams are displayed in a frequency versus time format. LOFARGRAMS contain information which, to the skilled operator, can assist in classification and analysis of contact motion. In Figure 8.7 the contact is strongest in beam 2; there has been a change in doppler, which indicates either a speed or a course change; and the frequencies of the tonals may classify the contact.

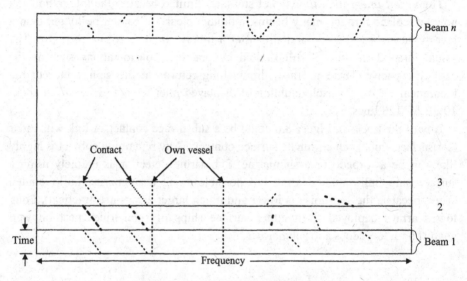

Figure 8.7   Narrowband (LOFAR) display

8.11   Formulae for Detection Threshold

The basic passive sonar equation (Chapter 7) may now be modified to include expressions for the *detection threshold* (DT) which are specific to the type of passive sonar. Formulae for DT are traditionally based on a square law detector acting on a signal in Gaussian noise, whose gain is given by

$$\left(\frac{\text{mean}(S+N) - \text{mean}(N)}{\text{std dev}(N)}\right)^2_{\text{out}} = BT\left(\frac{\text{signal power}}{\text{noise power}}\right)^2_{\text{in}}$$

which, from the definitions of d and DT given in Chapter 7, becomes

$$d = BT(DT)^2 \quad \text{or} \quad DT = 5 \log d - 5 \log BT \text{ (in dB form)}$$

The product BT measures the number of independent noise samples combined in the detector. Therefore the variance of the noise is reduced by BT relative to the mean, resulting in the gain given by this expression.

8.12   Broadband Square Law Detector

This passive system simply beamforms using the *full aperture* of the array and sums the energy in the bandwidth (usually split into separate octaves) of the expected signal over a period T_e, the electronic integration time. Therefore

$$DT = 5 \log d - 5 \log BT_e$$

and, substituting for DT, the sonar equation for this passive receiver, also called an *energy detector*, becomes

$$SE = (SL - PL) - N + DI - (5 \log d - 5 \log BT_e)$$

The expression for DT assumes a decision based on a single sample (or line for a practical multibeam receiver), displayed to the operator or passed to an automatic system. In practice, operators and automatic systems make decisions based on a number of successive time samples. The equation must therefore be modified by including a *visual integration term*, $5 \log n$, where n is the number of successive time samples used by the operator or automatic system in making the decision. Putting SE = 0 gives

$$\boxed{PL = SL - N + DI - 5 \log d + 5 \log BT_e + 5 \log n}$$

Note that T_e – the time that a signal may be expected to remain in a beam – is the time between successive display samples (lines) and is greater than $1/B$.

8.13   Broadband Cross-Correlator Detector

Here the receive array is split into two equal halves to form pairs of codirectional beams whose outputs are then complex cross-correlated. The phase outputs from this process provide bearing measurements to a resolution of typically one-tenth of the beamwidth (given adequate S/N ratio). Full beams can only improve on bearing resolution by using amplitude comparisons between adjacent beams, inherently inferior to phase comparisons.

The DT is now given by

$$\mathrm{DT} = 5\log d - 5\log 2BT_e$$

and the sonar equation becomes

$$\boxed{\mathrm{PL} = \mathrm{SL} - N + \mathrm{DI} - 5\log d + 5\log 2BT_e + 5\log n}$$

Note that the DI is now that of the *half-beams*, or 3 dB less than for the full aperture. However, the factor of 2 in the $5\log 2BT_e$ term implies a *net loss* of 1.5 dB (3 − 1.5) compared with the energy detector, a small price to pay for the improved bearing discrimination.

8.14 Narrowband Processor

For a narrowband signal processor, the square law detector must be modified to allow for the different bandwidths of signal and noise. For the *broadband* case

$$\left(\frac{S}{N}\right)^2_{out} = BT_e \left(\frac{S}{N}\right)^2_{in}$$

Here signal and noise apply to identical bandwidths, but for *narrowband* the signal is the intrinsic bandwidth of the line, whereas the noise is referred to a 1 Hz band. Both levels must be referred to a common bandwidth, the analysis bandwidth, B, of each frequency cell. The input noise to the narrowband processor is therefore narrowband, and for the narrowband case

$$\left(\frac{S}{N}\right)^2_{out} = BT_e \left(\frac{S}{NB}\right)^2_{in}$$

and taking logarithms, $DT = 5 \log d - 5 \log(T_e/B)$.

The *total processing time*, T_e, comprises two factors:

- The analysis time, $1/B$

- An integration factor, IF, which is the number of independent samples from the signal processor summed before display.

So $T_e = IF/B$; substituting for T_e we obtain

$$DT = 5 \log d - 5 \log \left(\frac{IF}{B^2}\right)$$

and simplifying gives

$$DT = 5 \log d + 10 \log B - 5 \log IF$$

If the *signal bandwidth*, B_s, is greater than the analysis bandwidth, B, the signal will be over resolved – it will appear in more than one cell – and there will be a reduction in gain of $10 \log(B/B_s)$. Note that any mismatch cannot increase processing gain and therefore any positive value for $10 \log(B/B_s)$ must be reduced to zero.

The expression for DT is then

$$DT = 5 \log d + 10 \log B - 10 \log(B/B_s) - 5 \log IF$$

and the *narrowband (power)* sonar equation is

$$\boxed{PL = SL - N + DI - 5 \log d - 10 \log B + 10 \log(B/B_s) + 5 \log IF + 5 \log n}$$

8.15   Narrowband Amplitude Detector Processor

If the signal processor produces amplitude values (usually by simply taking the square roots of the power values), the expression for DT becomes

$$DT = 10 \log \left[\frac{0.27d}{IF} + 1.05 \left(\frac{d}{IF} \right)^{1/2} \right] + 10 \log B - 10 \log(B/B_s)$$

and the *narrowband (amplitude)* sonar equation is

$$\boxed{\begin{array}{l} PL = SL - N + DI - 10 \log \left[\dfrac{0.27d}{IF} + 1.05 \left(\dfrac{d}{IF} \right)^{1/2} \right] \\[4mm] \qquad - 10 \log B + 10 \log(B/B_s) + 5 \log n \end{array}}$$

8.16 Worked Examples

Two passive sonar systems have been chosen to illustrate realistic performances against torpedoes and submarines: (1) submarine-mounted flank arrays of length 8 m and height 4 m, and (2) a surface ship towed array, 32λ at all frequencies by selecting elements. The 'threat table' (Table 8.2) will be used for both examples and gives realistic radiated spectrum noise levels for a noisy, thermally propelled torpedo (torpedo A); a quiet, electrically propelled torpedo (torpedo B); and a typical submarine (submarine C).

Table 8.2 Threat table for Examples 8.1 and 8.2

	Spectrum levels of radiated noise (dB)		
	Torpedo A	Torpedo B	Submarine C
Broadband signal (Hz)			
2000–4000	140	110	100
1000–2000	145	115	105
Narrowband signal (Hz)			
400	155	125	–
200	165	135	–
80	–	–	130
40	–	–	140

Example 8.1

The submarine passive sonar receives using flank arrays of length 8 m and height 4 m. If the design frequency of the array is 2000 Hz then $\lambda/2 = 375$ mm and the array will have 10 rows of 20 elements.

Assume the self-noise of the submarine to be equivalent to SS2. The background noise will then be ambient sea noise or SS2, whichever is the greater. The DI will be given by

$$\boxed{\text{DI} = 3 + 10\log mn + 20\log(f/f_0) \quad \text{(dB)}}$$

At 3000 Hz	$\text{DI} = 3 + 10\log 200 + 20\log 1.5$	$= 29$ dB
At 400 Hz	$\text{DI} = 3 + 23 - 20\log 5$	$= 12$ dB
At 200 Hz	$\text{DI} = 3 + 23 - 20\log 10$	$= 6$ dB
At 80 Hz	$\text{DI} = 3 + 23 - 20\log 25$	$= 3$ dB
At 40 Hz	$\text{DI} = 3 + 23 - 20\log 50$	$= 3$ dB

Because the flank arrays are baffled by the hull, DI cannot be less than 3 dB. The performance of the system is given by the following tables. The ranges have been calculated assuming spherical spreading plus absorption.

Narrowband

The narrowband equation is

$$\boxed{PL = SL - N + DI - 5\log d - 10\log B + 10\log(B/B_s) + 5\log IF + 5\log n}$$

and the system performance is given in Table 8.3.

Table 8.3   Narrowband system performance for Example 8.1

	Threat A		Threat B		Threat C	
Tonal	200	400	200	400	40	80
SL	165	155	135	125	140	130
N	73	68	73	68	85	80
DI	6	12	6	12	3	3
$5\log d$	10	10	10	10	10	10
B	2	4	2	4	0.1	0.2
B_s	4	8	4	8	0.2	0.1
$10\log B$	3	6	3	6	−10	−7
$10\log(B/B_s)$	−3	−3	−3	−3	−3	0
IF $(= BT_e)$	4	8	4	8	1	2
$5\log IF$	3	5	3	5	0	2
n	30	30	30	30	60	60
$5\log n$	7	7	7	7	9	9
PL	92	92	62	62	64	61
R (km)	40	40	1.3	1.3	1.6	1.1

Broadband

The broadband sonar equation is

$$\boxed{PL = SL - N + DI - 5\log d + 5\log BT_e + 5\log n}$$

and the system performance is given in Table 8.4.

Table 8.4 Broadband system performance for Example 8.1

	Threat A	Threat B	Threat C
SL	140	110	100
N	54	54	54
DI	29	29	29
$5 \log d$	6	6	6
B	2000	2000	2000
T_e	2	2	10
$5 \log BT_e$	18	18	21
n	30	30	60
$5 \log n$	7	7	9
PL $(\alpha = 0.2)$	134	104	99
R (km)	>100	50	40

Note that all three threats are *detected* by their broadband radiated noise at long range but, except for the noisy torpedo (threat A), *classification* by narrowband tonals is only possible at very close range. This is because, at the lower frequencies of the tonals, the background noise is much higher and the DI of the array is low (12 dB at best).

Ever quieter submarines and torpedoes are the stimulus behind the development of towed arrays, which can achieve a good DI down to the very low frequencies of the residual tonals of modern threats. The second example will show how, even when towed by a surface ship, towed arrays significantly improve classification ranges.

Example 8.2

The towed array will be 32λ at all frequencies by selecting elements. The background noise at the array will be the ambient sea noise or the radiated noise of the tow ship at the array, whichever is the greater. Note that a good towed array can be expected to achieve self-noise levels equivalent to SS2. Take the spectrum level of the tow ship radiated noise to be

- 140 dB from 40 to 400 Hz

- 130 dB from 1000 to 2000 Hz

The radiated noise at the array will be reduced by the separation from the ship (cable length) and by the main lobe to sidelobe ratio of the array. Take the cable length as 600 m and the ratio as −20 dB. Then, from 40 to 400 Hz, radiated noise at the array is

- $140 - 20 \log 600 - 20 = 64$ dB at operational array bearings
- $140 - 20 \log 600 - 0 = 84$ dB at ahead array bearings

And from 1000 to 2000 Hz, radiated noise at the array is

- $130 - 20 \log 600 - 20 = 54$ dB at operational array bearings
- $130 - 20 \log 600 - 0 = 74$ dB at ahead array bearings

Operational bearings for hull-mounted arrays are conventionally from $R135°$ around to $G135°$ (red for port, green for starboard), i.e., those bearings not directly affected by stern noise. By the same token, operational bearings for towed arrays are from $R45°$ around to $G45°$.

If the ambient sea noise is taken as SS2, the levels are as follows:

Hertz	40	80	200	400	1500
Decibels	85	80	73	68	58

The directivity index will be constant with frequency and given by

$$DI = 10 \log 64 \quad (64 \text{ elements spaced } \lambda/2)$$

$$= 18 \text{ dB at all frequencies from 40 to 2000 Hz}$$

For the array to be 32λ at 40 Hz, its length must be

$$\frac{32 \times 1500}{40} = 1200 \text{ m} \qquad (\text{using } c = f\lambda)$$

The performance of the system at operational bearings is given by the following tables. The ranges have been calculated assuming spherical spreading plus absorption. (Figure 3.3).

Narrowband

We have

$$\boxed{PL = SL - N + DI - 5 \log d - 10 \log B + 10 \log(B/B_s) + 5 \log IF + 5 \log n}$$

and the system performance is given in Table 8.5.

Table 8.5 Narrowband system performance for Example 8.2

	Threat A		Threat B		Threat C	
Tonal	200	400	200	400	40	80
SL	165	155	135	125	140	130
N	73	68	73	68	85	80
DI	18	18	18	18	18	18
$5 \log d$	10	10	10	10	10	10
B	2	4	2	4	0.1	0.2
B_s	4	8	4	8	0.2	0.1
$10 \log B$	3	6	3	6	−10	−7.
$10 \log(B/B_s)$	−3	−3	−3	−3	−3	0
IF $(= BT_e)$	4	8	4	8	1	2
$5 \log$ IF	3	5	3	5	0	2
n	30	30	30	30	60	60
$5 \log n$	7	7	7	7	9	9
PL	104	98	74	68	79	76
R (km)	160	80	5	2.5	9	6

Broadband

We have

$$PL = SL - N + DI - 5 \log d + 5 \log BT_e + 5 \log n$$

and the system performance is given in Table 8.6.

Table 8.6 Broadband system performance for Example 8.2

Threat	Threat A	Threat B	Threat C
SL	140	110	100
N	54	54	54
DI	18	18	18
$5 \log d$	6	6	6
B	1000	1000	1000
T_e	2	2	10
$5 \log BT_e$	16	16	20
n	30	30	60
$5 \log n$	7	7	9
PL	122	92	88
R (km)	>100	30	20

Note that all three threats are now *detected* by their broadband radiated noise at long range and also *classified* by narrowband tonals at useful ranges. This is because, even at the lower frequencies of the tonals, the DI of the array is maintained at 18 dB.

Over *operational bearings*, performance is limited by the ambient noise of the sea and therefore the performance would be the same if the tow vessel were a submarine.

8.17   Passive Ranging

When a passive sonar detects a signal from a radiating target, it gives a measure of the bearing of the target and how it changes with time. The range of the target, however, is not directly known; comparable signals at the array may result, for example, from a distant noisy source or from a close quiet source.

Several techniques are exploited by passive sonars to estimate the range of a target. They are particularly suitable for submarines whose bows and flanks provide low noise sites for large arrays. A submarine is a very difficult target to detect and therefore it will be reluctant to use active sonar, which immediately advertises its presence, to determine the range of a contact that may have already been detected and classified passively.

8.18 Triangulation

Given two well-separated arrays, the range can be estimated using simple trigonometry. This technique, called *triangulation*, is not limited to sonar and is best illustrated by a practical example. A submarine is in contact with a target at range R from a flank array and from a towed array whose centre is $S = 1000$ m behind the flank array. The maximum response angles (MRAs) are shown in Figure 8.8.

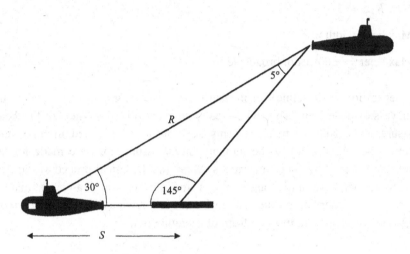

Figure 8.8   Ranging by triangulation

From simple trigonometry, given one side and two angles, we have

$$R = \frac{1000 \sin 145°}{\sin 5°} = 6581 \text{ m}$$

This estimate is highly dependent on the accuracies of the bearing measurements. If the beamwidths of both arrays are 4° then both MRAs could easily be in error by ±1° (one-quarter of the beamwidths), giving

$$R_{\text{max}} = \frac{1000 \sin 146°}{\sin 3°} = 10\,685 \text{ m}$$

$$R_{\text{min}} = \frac{1000 \sin 144°}{\sin 7°} = 4823 \text{ m}$$

The accuracy of the range estimate is proportional to the angle subtended at the target by the two arrays. Increasing the separation, S, of the arrays and/or reducing the range of the target increases this angle and improves the accuracy.

The arrays must be large enough for the beamwidths to be small at the frequencies of interest. Furthermore, increasing beamwidths as steer angles move towards endfire will limit the usefulness of the technique to angles not too far removed from broadside.

Practical limits to the technique might be

- Max $R/S = 5$

- Max beamwidth $= 5°$

- Max steer $= \pm60°$ from broadside

The separation (or baseline) for measurements may be greatly increased by using own vessel movement. This will greatly increase the baseline for the bearing measurements. But the measurements are now also separated in time, and in practice the target will also be moving. Initial assumptions are made for target course and speed, e.g., constant values. Many mathematical procedures are available to refine these initial assumptions and the resulting range estimate will improve with observation time. The technique is known as *bearings only analysis* (BOA) and is clearly an implicit form of triangulation.

8.19 Vertical Direct Passive Ranging

Measurements of the vertical angles of arrival at an array of a signal from a radiating target and the time differences between signals reaching the array by different paths can be used to estimate the range and depth of the target (Figure 8.9). The technique is known as *vertical direct passive ranging* (VDPR).

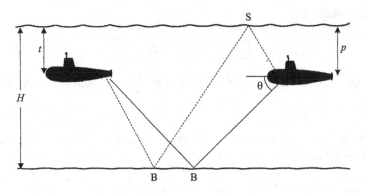

Figure 8.9 VDPR range estimation

Measuring the vertical angles of arrival to an adequate accuracy requires an array with a good vertical aperture at all relevant frequencies. *Flank and bow arrays* on submarines can therefore provide a complete VDPR facility but *towed arrays* with *no* vertical directivity are restricted to measuring time differences by autocorrelation only. The expressions used rely on the following assumptions:

- Platform and target depths, p and t, are much less than the water depth, H
- The sea bed is flat at the bounce site

The range of the target in terms of arrival angle is given by the following approximate expressions:

- Bottom bounce only (B, solid path)

$$R = (2H - p - t)/\tan \theta$$

- Bottom bounce and surface bounce (BS, dashed path)

$$R = (2H + p - t)/\tan\theta$$

- SB (not shown)

$$R = (2H - p + t)/\tan\theta$$

- SBS (not shown)

$$R = (2H + p + t)/\tan\theta$$

The vertical maximum response angle, θ, can be estimated by amplitude or by phase:

- Amplitude comparisons across a vertical fan of full beams (formed using complete staves)

- Phase comparisons between vertical half-beams (formed using top and bottom halves of the vertical staves)

Example 8.3

Suppose the signal is contained within an octave from 1000 to 2000 Hz and the height of the array is at least 5 wavelengths over this band, then from Figure 2.4, the vertical beamwidth for steers up to $\pm 30°$ will be at most about 12°. If we assume amplitude or phase comparisons will improve on this by a factor of 4, then measurements of the vertical arrival angle will be up to $\pm 3°$ in error.

For a simple bottom bounce only, $H = 1000$ m, $p = 100$ m, $t = 0$ (surface ship) and $\theta = 30°$, the range is $R = 3290$ m. The limits (for $\theta = 27°$ and 33°) will be 3728 m and 2925 m, or approximately $\pm 12\%$.

This might be seen as an acceptable result for a passive ranging system. But it does depend fundamentally on the accuracy of θ, which in turn implies a narrow vertical beamwidth. Unfortunately, detection – which must of course precede ranging – is increasingly only likely at low frequencies where beamwidths will be large. Therefore ranging using θ only is likely to be very inaccurate.

The *time difference of arrival*, δT, between the two paths is given approximately by

$$\delta T = \frac{2p\sin\theta}{c}$$

and for this example $\delta T = 67$ ms. δT is now used as an initial value for the offset between two 'windows' of time series data from the two paths shown. Cross-correlation is performed between the two time series and the correlation peak is used to improve on the initial value of 67 ms.

Suppose the improved value is 80 ms; using the above formula, $\theta = 37°$ and the range can now be determined from the earlier formulae:

- For the downward-looking beam $\quad R = 1900/\tan 37° = 2521$ m
- For the upward-looking beam $\quad R = 2100/\tan 37° = 2787$ m

The average value is 2654 m.

There is an alternative method which can also yield an estimate of target depth. Simple geometry in the vertical plane gives the following expressions for the sound transit times for the four possible paths that include one bottom bounce:

$$T_B = \frac{1}{c}\left[R^2 + (2H - p - t)^2\right]^{1/2}$$

$$T_{BS} = \frac{1}{c}\left[R^2 + (2H + p - t)^2\right]^{1/2}$$

$$T_{SB} = \frac{1}{c}\left[R^2 + (2H - p + t)^2\right]^{1/2}$$

$$T_{SBS} = \frac{1}{c}\left[R^2 + (2H + p + t)\right]^{1/2}$$

and for this example, provided p and t are small compared with H, then $\delta T = T_{BS} - T_B$ is approximately

$$\delta T = \frac{4Hp}{c(R^2 + 4H^2)^{1/2}}$$

If $\delta T = 80$ ms then for this example $R = 2667$ m. This is very close indeed to the average value obtained above (2654 m), by simply using the angle derived from the measured δT in the coarse range formulae.

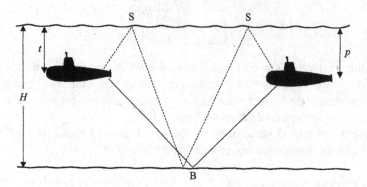

Figure 8.10   VDPR depth estimation

Having obtained the range from one time difference, the depth of the target can be estimated from the time difference between two other paths. If we take the difference between the two paths in Figure 8.10, $\delta T = T_{SBS} - T_B$, then we obtain

$$\delta T = \frac{4H(p + t)}{c(R^2 + 4H^2)^{1/2}}$$

Suppose $\delta T = 200$ ms and the range, already obtained from different paths, is 2667 m then $t = 150$ m.

8.20   Horizontal Direct Passive Ranging

The curvature of the wavefront of a signal from a distant target determined by measurements at three arrays can be used to estimate the range of the target (Figure 8.11). The technique is known as *horizontal direct passive ranging* (HDPR).

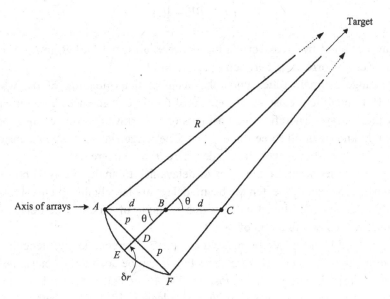

Figure 8.11   HDPR geometry

Consider three collinear equispaced arrays *A*, *B* and *C*. The target is at range *R* and at an angle θ with the axis of the arrays:

$$R^2 = p^2 + (R - \delta r)^2$$

where δr is the additional path length due to curvature. We have $p = d \sin \theta$, therefore

$$R^2 = (d \sin \theta)^2 + (R - \delta r)^2$$

and when *R* is large compared with *d*, we have

$$\boxed{R = \frac{(d \sin \theta)^2}{2\,\delta r}}$$

To find the range, we therefore need to determine δr and θ. For a distant target, angle θ is almost the same for all three arrays. In practice, determine θ by measuring and averaging the maximum response angles from the three arrays; δr is given by

$$\delta r = BE - \tfrac{1}{2}CF$$

BE can be found by cross-correlating between arrays A and B, and CF can be found by cross-correlating between arrays A and C.

The range estimate relies upon the accurate determination of δr, which is dependent upon the precise measurement of the time differences between arrivals at the three arrays. Therefore the positions of the arrays themselves must be very accurately known in all three dimensions. The range estimate is less sensitive to bearing errors, particularly at broadside where $\theta = 90$ degrees.

The delays measured are the sums of delays due to angle of arrival plus delays due to wavefront curvature. On the beam, delays are largely due to curvature; close to endfire, delays are almost entirely due to angle of arrival. The delays due to curvature reduce as a function of $\sin^2 \theta$.

The method is optimized by making δr – the difference in path lengths due to curvature only – as large as possible so that it can be accurately determined. This occurs where d is large and is close to 90°. In practice, δr needs to be some significant proportion of a wavelength, say at least $\lambda/10$, for an accurate measurement, and for realistic values of d and array size, this limits the use of the technique to quite high frequencies.

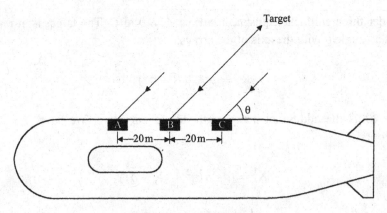

Figure 8.12   HDPR example

Example 8.4

Three arrays, A, B and C, are mounted on the flank of a submarine (Figure 8.12). Signals from a target at 30° to the axis of the arrays are detected in an octave centred on 3000 Hz at all three arrays. The delays measured by cross-correlation are

- Delay between arrays A and $B = 0.0067$ s

- Delay between arrays A and $C = 0.0133$ s

Referring to Figure 8.11, we have

$$BE = 0.0067 \, c = 10.05 \text{ m}$$

$$CF = 0.0133 \, c = 20 \text{ m}$$

$$\delta r = 10.05 - 10 = 0.05 \text{ m}$$

$$R = (20 \sin 30°)^2 / (2 \times 0.05) = 1000 \text{ m}$$

At 3000 Hz, the value of λ is 0.5 m, so δr is a small but significant part of a wavelength ($\lambda/10$). If the target were actually ±3° from its measured bearing, the true ranges would be 1186 m or 824 m. (These are very similar errors as for VDPR.)

For a target on the beam ($\theta = 90°$) then, for the same δr, we have $R = (20 \sin 90°)^2 / (2 \times 0.05) = 4000$ m. Also, for the same bearing measurement errors, the actual range would be 3988 m, i.e., a very small range error. Therefore, not only is δr largest for the same range when the target is on the beam, but also range accuracy is least sensitive to bearing errors. The submarine can optimize the range estimate by manoeuvring to put and keep the target on a bearing close to broadside.

In practice the arrays will not be equispaced and neither will they be aligned in the remaining two axes. The full equations for range and bearing of a target will therefore be three-dimensional and more complex, although based on the same simple geometric relationships. The simple case for collinear, equispaced arrays has clearly demonstrated the principle of HDPR and how a range estimate may be optimized.

8.21   Towed Arrays

Towed arrays are essential for the detection of modern, quiet submarines and torpedoes. This section discusses their construction, design and problems. A simplified diagram of a towed array is shown in Figure 8.13.

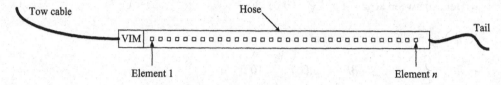

Figure 8.13   Simplified towed array

A typical towed array will comprise several hundred hydrophones, together with electronic circuits to preamplify, sample and digitize their outputs. The hydrophones are enclosed in a plastic *hose*, liquid filled to achieve neutral buoyancy in the sea. The array is towed by a *cable* of a length decided by the speed of the tow vessel and the desired depth of operation; it will be several hundreds of metres. A *vibration isolation module* (VIM) reduces the tow vessel vibrations transmitted down the cable, which would otherwise augment the noise of the array. The assembly is completed by a *tail* whose purpose is to provide some drag to the array so that it is maintained reasonably linear.

In practice the array will have significant *curvature*, which would affect the beam shapes unless corrected in the beamformer. *Heading sensors* are placed at intervals within the array and their readings used to correct for array curvature; there will be at least three sensors (front, rear and middle) and possibly more in very long arrays.

Towed arrays can have diameters as small as, say, 50 mm (thin arrays) and as large as, say, 150 mm. The length of a towed array is determined by its frequency of operation and the desired DI. When used as a receive array in an active system, the length might be from 10 to 50 m, whereas for a passive array the length might be from 100 to 1000 m.

Passive towed arrays operate over a frequency range of several octaves. The spacing between the elements is maintained at about $\lambda/2$ at the centre frequency of each octave by a suitable choice among the available elements.

Example 8.5

Design a towed array covering 200 to 1600 Hz with a DI of about 16 dB at all frequencies within this range.

$DI = 10 \log n = 16$. Therefore $n = 40$ and each octave must use 40 elements spaced $\lambda/2$ at its centre frequency (Table 8.7). The complete array is shown in Figure 8.14. The higher octaves are *nested* (a term possibly borrowed from computer programming) within the lower octaves, and elements are chosen to meet the requirements of each octave. To gain 3 dB in DI, or to operate at the next lower octave, the array must be doubled in length.

Table 8.7 Design parameters for Example 8.5

Octave (Hz)	f_0 (Hz)	$\lambda/2$ (m)	n	Length (m)
200–400	300	2.5	40	100
400–800	600	1.25	40	50
800–1600	1200	0.625	40	25

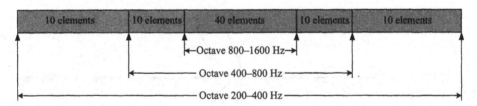

Figure 8.14 Nested array

8.22   Bearing Ambiguity

A single line array is omnidirectional in the vertical plane and therefore when horizontal beams are formed, they exhibit a *left/right ambiguity*. Figure 8.15 (top) illustrates the source of the left/right ambiguity. There are several methods available to resolve this ambiguity.

Course change

If the tow vessel, and therefore the towed array, changes its heading, it is possible to resolve the ambiguity as shown in Figure 8.15. The array heading change is not instantaneous and, particularly for a very long array, can take a considerable time. Nor is the target stationary. Nevertheless, the true target bearing is often quickly resolved, but may need confirming by making some assumptions about the target's motion.

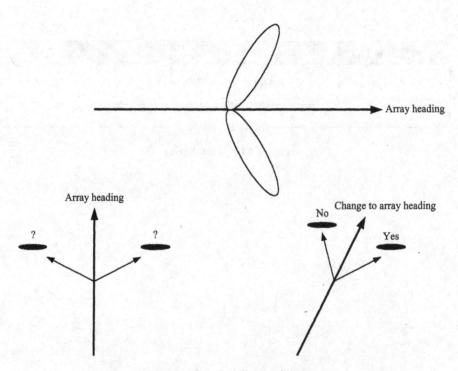

Figure 8.15   Left/right ambiguity

Twin arrays

Parallel twin towed arrays use the time delays between the signals arriving at the two arrays to resolve the left/right ambiguity. Maintaining the spacing between two flexible arrays is a practical problem, particularly for very long arrays and during a change of course. Precision, however, is not necessary provided some horizontal spacing survives and the arrays do not cross over. If both arrays are also used to form beams, an improvement in DI of up to 3 dB will result over a limited frequency range (particularly useful when the twin array is used as a receive array in an active system which will have a comparatively small percentage bandwidth).

Triplets

In a triplet array, each element now comprises three hydrophones – a triplet – in the vertical plane (Figure 8.16). Because time delays are measured between all three pairs, the left/right ambiguity can be resolved regardless of any rotation of the array. The method needs a fairly large array diameter to house the triplets and to produce measurable time delays.

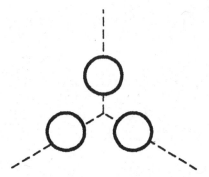

Figure 8.16   Hydrophone triplet

8.23   Self-noise

Towed arrays are well separated from tow vessels and therefore the vessel radiated noise is significantly reduced (given a cable length of 500 m, the value of $20 \log R$ is 54 dB) and, except for the ahead bearings of the towed array, this is further reduced by the main lobe to sidelobe ratio of the beams (perhaps 20 dB). The hydrodynamic noise of the towed array can be made negligible at normal tow speeds of up to about 12 knots, and therefore the remaining and dominant noise is the ambient noise of the sea.

8.24 Problems

8.1 The dimensions of the flank arrays of Example 8.1 are halved. How are the narrowband and broadband performances affected?

8.2 Refer to Figure 8.9. The platform depth, p, is 300 m, the target depth, t, is unknown but small compared to the water depth, H, of 2000 m. If the time difference between the paths shown is 100 ms, what is the range of the target?

8.3 Design a towed array to have a DI of 21 dB at the centre frequency of the octave from 2000 to 4000 Hz. What would be the spectrum level of the radiated noise of a target which is first detected at a range of 10 km against a background noise equivalent to SS4? Assume broadband detection and $5 \log d = 6$ dB, $T_e = 10$ s, $n = 60$. Propagation is spherical spreading plus absorption.

9
Active Sonar

9.1 Range, Pings and Doppler Shift

Active sonars use an array of projectors to transmit acoustic pulses into the water. Underwater targets are detected, localized and classified by the echoes resulting from their insonification by these pulses.

The time of echo arrival at the receiver array is used to determine the range of a target (strictly, the slant range). The time, measured from transmission, is the time taken for the sound to propagate to the target and back to the array. Hence the range is given by

$$\boxed{R = ct/2}$$

An active sonar transmission is known as a *ping*. A ping may be simply the transmitted pulse or sequence of pulses, or it may be the total time between transmissions, i.e., the sum of the duration of the pulses and the receive period. The term 'ping' is therefore ambiguous but the meaning is usually made clear by the context in which it is used.

The *ping interval* is an alternative term for the time between transmissions and is the reciprocal of the ping (or pulse) repetition rate (PRR). The frequency shift of an echo is a measure of the *relative velocity* (doppler) of a target with respect to the sonar platform. The total shift is the sum of four components (Figure 9.1):

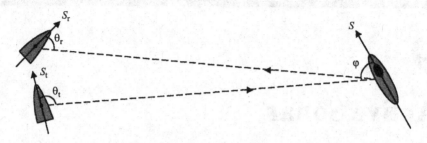

Figure 9.1 Components of doppler shift

- A doppler shift to the transmitted pulse, equal to $S_t \cos \theta_t$
- A doppler shift to the pulse received by the target, $S \cos \varphi$
- An equal doppler shift to the echo transmitted by the target, $S \cos \varphi$
- A doppler shift to the echo received by the platform, $S_r \cos \theta_r$

The first and fourth components are normally removed by *own doppler nullification* (ODN). The total frequency shift due to the target doppler, $S \cos \varphi$, is given by

$$\Delta f = \frac{2S \cos \varphi}{f/c}$$

where f is the operating frequency. The factor of 2 results from the two equal doppler shifts due to target motion (the second and third components).

Practical formulae for doppler shift are, for $c = 1500$ m/ s,

$$\Delta f = \pm 0.69 \text{ Hz per knot per kHz}$$
$$\Delta f = \pm 1.33 \text{ Hz per m/s per kHz}$$

Note that 1 m/s $= 1.945$ knots. And remember that doppler is a shift of *relative velocity* not speed. For a passive sonar, only the third and fourth components are relevant. The frequency shift is *positive* for an *approaching* target but may be changed by heterodyning in the receiver.

9.2 Pulse Types

Active sonars know more about the signal to be detected, and therefore the receiver is designed to match the signal, i.e., it uses *matched filter processing*. But the background against which the signal has to be detected, contains *reverberation* in addition to the ambient and self-noise of passive sonars. This additional background is all important in active systems and the designer must be aware of its magnitude and how to discriminate against it.

Active sonars employ two broad classes of pulse types:

- *Continuous wave (CW)*: a pulse of constant frequency and duration T seconds. The bandwidth of the pulse, and of the matched filter for optimum detection of this pulse, is $1/T$ Hz.

- *Frequency modulation (FM)*: the frequency of the pulse changes during the T seconds duration of the pulse. The bandwidth, B, is *not* now the inverse of the pulse length.

The replica correlation process used to detect an FM pulse is effectively a matched filter. Independent outputs from the correlator occur every $1/B$ seconds, i.e., at the *resolved pulse length*.

9.3 CW Processing

CW processing is similar to narrowband passive processing where, after beam-forming, the data is analysed using a fast Fourier transform (FFT) process and then displayed to an operator and/or input to an automatic detection and classification system. Traditionally, active CW processing used a bank of analogue filters, each matched to the bandwidth of the CW pulse. Sufficient filters were provided to cover the expected frequency shifts due to doppler. In modern sonars the analogue filters are replaced by an FFT processor.

9.4 FM Processing

FM processing replaces the FFT by a process where the output of the beamformer is correlated with a replica of the transmitted pulse. The signal processing for both classes of pulse is essentially *matched filter processing* and the same expression for signal processing gain therefore applies to both.

The signal-to-noise gain from replica correlation arises from a reduction in the mean of the noise because, unlike the signal, it is uncorrelated with the replica pulse shape. The DT equation assumes that the noise is completely uncorrelated with the transmitted pulse and therefore the noise power is reduced by the number of independent samples combined in the correlation process, which is BT. Thus

$$\boxed{\frac{\text{mean noise power}}{\text{after correlation}} = \frac{1}{BT}\left(\frac{\text{mean noise power}}{\text{before correlation}}\right)}$$

and the DT equation from this process is

$$\boxed{DT = 5\log d - 10\log BT}$$

Note that the gain is $10\log BT$ compared with $5\log BT$ for broadband passive sonars. This is because replica correlation is a coherent process (complete knowledge of the signal) whereas broadband processes are incoherent (nothing known of the signal).

- A coherent gain is equal to the number of independent samples $= BT$

- An incoherent gain is equal to the square root of the number of independent samples $= (BT)^{1/2}$

9.5 Active Sonar Equations

There are two active sonar equations; one is used to determine the performance against a background of *noise* and the other against a background of *reverberation*. Although some models combine the two backgrounds to estimate performance, it is preferable to calculate noise and reverberation performance separately for a better understanding of the effects of changing equipment parameters and operating in different environments.

The *noise-limited* active sonar equation is

$$SE = SL + TS - 2PL - (N - DI + 10 \log B) - (5 \log d - 10 \log BT - 5 \log n)$$

where $(N - DI + 10 \log B)$ is the *in-beam noise* over the full bandwidth and $(5 \log d - 10 \log BT - 5 \log n)$ is the *detection threshold*. Here n is the number of pings used by the operator or automatic system to make the decision; in a passive system, n is the number of lines used.

Putting $SE = 0$ and combining terms:

$$\boxed{2PL = SL + TS - N + DI + 10 \log T - 5 \log d + 5 \log n}$$

It is important to note that the noise-limited performance is independent of bandwidth (given matched filter processing). Therefore, FM and CW transmissions of equal duration have identical noise-limited performance.

The above equation was derived assuming perfect matched filter processing and may require modification as follows:

CW pulse: If the analysis bandwidth is greater than the pulse bandwidth there will be a loss (due to the increase in noise): If the analysis bandwidth is smaller than the pulse bandwidth there will also be a loss (due to the loss of some of the signal). The loss is given by a mismatch term:

$$\text{either, } -10 \log(B_a/B_s) \text{ when } B_a > B_s \text{ or, } -10 \log(B_s/B_a) \text{ when } B_a < B_s$$

LPFM Pulse: If the replica is extended (to allow for target Doppler) there will be a constant loss (due to the increase in noise) for *all dopplers* of

$$-10 \log(B_r/B_s)$$

If the replica is not extended but there is a mismatch (an incomplete overlap) with the signal return due to target doppler, there is a variable loss given by

$$-20 \log(B_r/B_c)$$

where B_c is that part of the signal which overlaps the reference. Note that these mismatch terms are always negative, so that – when added to the RHS of the equation – they always reduce 2PL.

The *reverberation-limited* active sonar equation is

$$\text{SE} = \text{SL} + \text{TS} - 2\text{PL} - (\text{SL} - 2\text{PL}_R + \text{TS}_R) - (5 \log d - 10 \log BT - 5 \log n)$$

where $(\text{SL} - 2\text{PL}_R + \text{TS}_R)$ is the *in-beam reverberation* over the full bandwidth and $(5 \log d - 10 \log BT - 5 \log n)$ is the *detection threshold*.

When the propagation paths for echo and reverberation are identical, $2\text{PL} = 2\text{PL}_R$ and these terms will cancel out.

TS_R, the target strength for reverberation, is given by

$$\text{TS}_R = S_b + 10 \log A$$

The dominant reverberation source is usually at the sea surface, the sea bed or in a horizontal layer within the water column (see Chapter 6) and therefore

$$\text{TS}_R = S_b + 10 \log \frac{cT}{2} \left(\frac{2\pi}{360} \right) R\theta_h$$

and putting SE = 0 gives

$$10 \log R = 10 \log(1/T\theta_h) - S_b - 41 + \text{TS} - 5 \log d + 10 \log BT + 5 \log n$$

and combining terms:

$$\boxed{10 \log R = 10 \log(B/\theta_h) - S_b - 41 + \text{TS} - 5 \log d + 5 \log n}$$

Note that the 41 dB arises from

$$10 \log(c\pi/360) + 30 \text{ (to give } R \text{ in km)} = 11 + 30 = 41 \text{ dB}$$

The term $10\log(B/\theta_h)$ is known as the *reverberation index* (RI) and is a useful measure for comparing performance against a background of reverberation.

For a *CW pulse*, $B = 1/T$, where T is the pulse length.

For an *FM pulse*, $B = 1/T_r$, where T_r is the *resolved pulse length*, or the time between independent output samples from the correlator.

Example 9.1

CW pulse: $\qquad\qquad T = 100$ ms, $\theta_h = 10°$

$$RI = 10\log(10/10) = 0 \text{ dB}$$

Example 9.2

FM pulse: $\qquad\qquad B = 300$ Hz $\quad \theta_h = 10°$

The actual pulse length is irrelevant and $RI = 10\log(300/10) = 15$ dB.

The wider bandwidth of the FM pulse therefore results in a much improved performance against reverberation. Looking at it another way, the effective reverberating area is determined by the bandwidth and not by the pulse length.

If the dominant factor is *volume reverberation* resulting from backscattering from that part of the water column which is actually included in *both* transmit and receive beam patterns, then the equation becomes

$$20\log R = 10\log(B/\theta_h.\theta_v) - S_v - 23 + TS - 5\log d + 5\log n$$

Note that the 23 dB arises from

$$10\log\left(\frac{c\,\pi}{2\,4}\frac{\pi^2}{180^2}\right) + 30 \text{ (to give } R \text{ in km)} = -7 + 30 = 23 \text{ dB}$$

9.6   Reverberation Index

The reverberation index is a measure of the effectiveness of the sonar system against reverberation. The index may be increased for *all* pulse types by reducing the effective horizontal beamwidth. The *receive* array must therefore be large (many wavelengths) in the horizontal dimension.

Increasing the pulse bandwidth, B, will also increase RI but, for a simple CW pulse, where increasing the bandwidth is only possible by reducing the pulse length, T, the noise-limited performance is adversely affected. This, however, had to be the approach in early sonars where the complexities of FM were beyond the available technology. Long and short CW pulses would be used in the same ping. Modern sonars employ two standard pulse design approaches to improve performance against reverberation:

- A wideband FM pulse which spreads the reverberation power over the bandwidth, B, of the pulse. When the signal is correlated with a replica of the transmitted pulse – effectively coherent matched filter processing – the pulse is compressed into an equivalent time – *the resolved pulse length, T_r*, which is equal to the inverse of B. The area of the surface contributing reverberation is therefore reduced and performance against reverberation is improved as B is increased, increasing B in $10\log(B/\theta_h)$. Furthermore, the performance of the wideband pulse is independent of target doppler – provided there is still an adequate match with a replica.

- A long, shaped CW pulse is used to produce a reverberation spectrum with a narrow peak centred on zero doppler whose amplitude falls off rapidly with increasing magnitude of doppler. Target doppler will ensure that target echoes fall into regions of the reverberation spectrum where the reverberation power is low and the performance eventually becomes *noise limited only*.

An alternative broadband pulse is the *pseudorandom noise* (PRN) pulse. Its bandwidth will be close to or identical to that of an FM pulse designed for a similar performance. The frequency–time structure of the pulse may be changed in a random but known manner from ping to ping. Hence replica correlation will still be possible.

9.7   FM pulses

In the strict radio communications sense, sonar FM pulses are not frequency modulated at all. They are simply wideband pulses where the frequency changes throughout the duration, T, of the pulse (Figure 9.2).

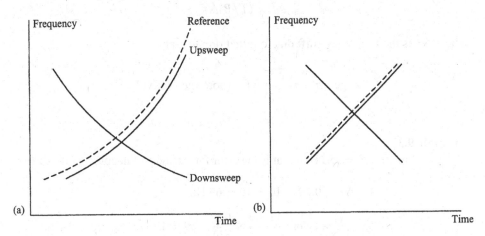

Figure 9.2   Sonar pulses: (a) LPM and (b) LFM

Linear period FM (or hyperbolic FM)

The periodicity of the pulse changes linearly with time (i.e., its frequency changes hyperbolically with time). Therefore the echoes, shifted in frequency by target doppler, will always correlate with a suitably extended replica. The extended replica will result in a constant loss, perhaps as much as 2 or 3 dB, for the entire range of target doppler and it is often preferable to use a replica equal to the pulse bandwidth and to accept a loss increasing with target doppler. No doppler information can be extracted from this pulse.

Linear FM

The frequency of the pulse changes linearly with time. The echoes shifted by target doppler will not now fully correlate with an extended replica and a set of replicas (analogous to the comb of filters or FFT cells used to process CW echoes) are now required to cover the expected target dopplers. The best replica match indicates the target doppler.

Range error with target doppler

Target doppler moves the instant in time when full correlation of the echo and replica takes place. There will only be an error when a single replica is used (the LPFM case). To a first order, the time error is

$$\Delta t = (T/B)\Delta f$$

where Δf is the frequency shift due to target doppler, and

$$\text{Range error} = \frac{c}{2}\frac{T}{B}\Delta f \quad \text{(note the 'active' 2)}$$

Example 9.3
If $T = 500$ ms and $B = 250$ Hz, what is the error for 10 knots of doppler at 10 kHz?

$$\Delta f = 0.69 \times 10 \times 10 = 69 \text{ Hz}$$

$$\text{Range error} = \frac{1500}{2} \times \frac{0.5}{250} \times 69 = 104 \text{ m}$$

The target doppler may be determined by range rate calculations from two or more pings or from a simultaneous CW transmission.

9.8   CW Pulses

A *rectangular* shaped pulse is defined in the time domain as

$$W_{R(t)} = \begin{cases} A \sin 2\pi\, ft & \text{for } |t| \leq T/2 \\ 0 & \text{otherwise} \end{cases}$$

and the frequency spectrum of the pulse, normalized to 1 s and 1 kHz, is shown in Figure 9.3. The 3 dB bandwidth is given by

$$\boxed{\Delta f_3 = \frac{0.91}{T}}$$

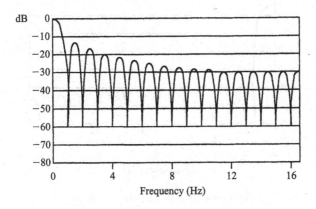

Figure 9.3   Rectangular pulse spectrum

A *Hamming* shaped pulse is defined in the time domain as

$$W_{H(t)} = \begin{cases} A[0.54 + 0.46\cos(2\pi t/T)]\sin 2\pi\,ft & \text{for } |t| \leqslant T/2 \\ 0 & \text{otherwise} \end{cases}$$

and again its frequency spectrum, normalized to 1 s and 1 kHz, is shown in Figure 9.4.

The 3 dB and 40 dB beamwidths are given by

$$\boxed{\Delta f_3 = \frac{1.37}{T} \qquad \Delta f_{40} = \frac{3.5}{T}}$$

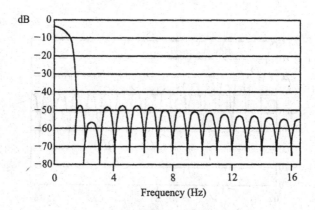

Figure 9.4 Hamming pulse spectrum

Note the significant reduction in sidelobe levels which is achieved, but at the expense of a 50 per cent increase in the 3 dB bandwidth.

Figures 9.3 and 9.4 are obtained by analysis using a rectangular window. If pulse and window have the same length, it is possible to exchange pulse and window weighting functions without changing the final observed pulse spectrum. In other words:

- A shaped pulse analysed using a rectangular window has the same spectrum as a rectangular pulse analysed using a shaped window.

The frequency spectrum of a *Hamming* shaped pulse analysed using a *Hamming* shaded window, again normalized to 1 s and 1 kHz, is shown in Figure 9.5. The 3 dB and 40 dB bandwidths are given by

$$\Delta f_3 = \frac{1.76}{T} \qquad \Delta f_{40} = \frac{5.5}{T}$$

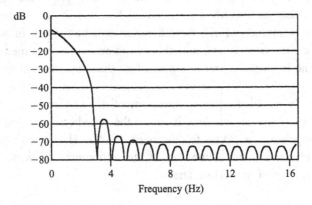

Figure 9.5   Hamming pulse: hamming window spectrum

A further reduction in sidelobe levels is achieved, but the 3 dB bandwidth is now almost doubled compared to the equivalent rectangular pulse analysed using a rectangular window.

All these spectra are theoretical. The need to tune projectors will 'knock the corners' off a rectangular pulse, making the sidelobe levels lower than predicted; and the limitations of electronics and digital processing will make it impossible to achieve the very low sidelobe levels predicted for shaped pulses. It would be unwise to expect sidelobes more than about 40 dB down on the main lobe.

9.9   Reverberation Rejection by CW Pulses

As target doppler increases, detections take place against a reduced background
of reverberation, finally becoming noise limited. To reduce the possibility of
detection, a submarine will always attempt to minimize the magnitude of its
doppler as seen by a sonar receiver.

To achieve a good performance against reverberation, therefore, the CW pulse
must be designed to maximize *reverberation rejection*, R_j, at the target dopplers
of interest. Pulse bandwidth and sidelobe levels are the parameters which
determine the magnitude of R_j. Reverberation will be present in both the main
lobe and sidelobes of a beam, and this reverberation is modified by platform
movement. The importance and interplay of all these factors will now be considered.

Figure 9.6 shows the frequency spectrum of a 1 s Hamming pulse analysed
using a Hamming shaded window. It shows the reverberation rejection resulting
from target doppler for a carrier frequency of 10 kHz. The dotted envelope is that
of a rectangular pulse and demonstrates the importance of pulse and window
weighting to improve CW performance.

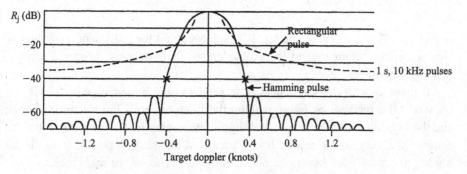

Figure 9.6   Reverberation rejection, R_j versus target doppler

For example, -40 dB of R_j is reached for 0.4 knots of target doppler using a
1 s Hamming pulse and window, but this is never reached using a rectangular
pulse for realistic submarine dopplers.

Figure 9.6 portrays a stable and stationary platform in a stable environment. If
this is not the case, the background reverberation in the main lobe and sidelobes of
a beam will be modified as follows.

9.10   Reverberation and Target Echoes in the Main Lobe

Beamwidth

It is only possible to fully nullify own doppler at one bearing within a beam. The finite beamwidth has two effects:

- It increases the doppler spread of the reverberation within a beam. The spreading is independent of pulse length and sea state. Echoes therefore need more doppler to avoid the reverberation background.

- There will be a related uncertainty in the measured target doppler.

If we assume a beamwidth of 10°, the spreading and doppler errors in knots at various relative bearings are as in Table 9.1

Table 9.1   Doppler spreads/errors due to beamwidth

Doppler spreads/errors (knots)			Platform speed (knots)
Target on bow	Target at 45°	Target on beam	
±0	±0	±0	0
±0.04	±0.6	±0.9	10
±0.06	±0.9	±1.4	15
±0.08	±1.2	±1.8	20

Ship motion

Roll and pitch have little effect on reverberation spread but contributions due to *yaw* are significant. NWS 1000 (UK Naval Weapons Specifications Publication) gives a figure for yaw in large frigates and destroyers as $1.75° \, s^{-2}$ peak. (Yaw is unlikely to be important in submarines or for towed arrays, except perhaps for very long pulses.) If the array is 60 m from the centre of gravity then the linear acceleration at a *bow dome* is $2 \, m/s^2$ and, by inference from other NWS 1000 figures, the doppler spreads for 1 s and 250 ms pulses are as in Table 9.2. These spreads occur both at transmit and receive.

Table 9.2 Doppler spreads/errors due to ship motion

Sea state	Pulse length (ms)	Doppler spreads/errors (knots)		
		Target on bow	Target at 45°	Target on beam
6	1000	0	1.8	2.7
4	1000	0	0.9	1.3
6	250	0	0.4	0.6
4	250	0	0.2	0.3

Environment

In *shallow water*, bottom reverberation can produce spreads of between 0.2 and 0.5 knots. In *deep water*, surface reverberation is important and depends on wind speed. A useful empirical formula is

$$\boxed{\text{Doppler spread} = 0.07v + 0.3}$$

with doppler and wind speed, v, in knots. And, substituting for the appropriate wind speeds, we obtain Table 9.3.

Table 9.3 Doppler spread due to sea state

Sea state	2	4	6
Doppler spread (knots)	1.2	1.6	2.3

Total reverberation

The total reverberation spread in the *main lobes* results from the pulse spectrum plus the three factors we have looked at:

- Beamwidth

- Ship motion

- Environment

Their relative importance is highly dependent on the sonar platform. Beamwidth and ship motion are irrelevant to a stationary platform, such as a sonobuoy or a helicopter dipping sonar, but the environment affects all platforms. Combining all the factors can result in quite large, worst case, values for reverberation spread. The likelihood of all factors coinciding, however, is not great and recommended practical figures for a first pass at a problem are as follows:

- Sonobuoy or helicopter $\pm$0.5 knot

- Towed arrays (ship or submarine) $\pm$0.7 knot

- Hull-mounted sonar, submarine $\pm$1.0 knot

- Hull mounted sonar, surface ship $\pm$1.5 knots

9.11   Reverberation and Target Echoes in the Sidelobes

The reverberation (including target echoes) which enters a receiver through the sidelobes of a beam is modified by an ODN frequency appropriate to the main lobe axis, and for full-beam processing this produces bands of reverberation, reduced only by the main lobe to sidelobe ratio (20–30 dB) appearing in the main lobe. This is a performance degradation apparent in many beams, or given an omnitransmission, in all beams; it can produce bands of reverberation which can affect the detection of targets with dopplers up to twice platform speed.

Figure 9.7 shows the mechanism of spreading due to platform speed. For a target in the *ahead beam* and a platform speed of v knots, both reverberation and target echoes received in the ahead beam are shifted by a frequency equivalent to v knots. An ODN frequency equivalent to v knots restores the reverberation in the ahead beam entering through the main lobe to a spectrum centred on zero knots. The reverberation entering through the *sidelobes* of the ahead beam, however, is also shifted by v knots and therefore the ahead beam output has a series of secondary reverberation spectra reduced by the main lobe to sidelobe ratios and centred on various speeds around to $-2v$ knots for reverberation entering through the astern sidelobe.

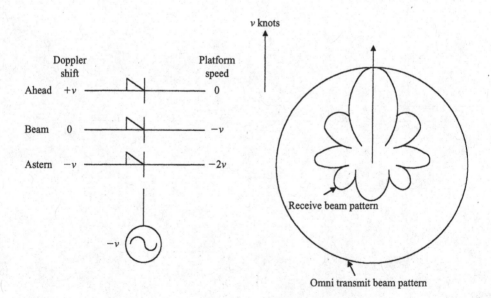

Figure 9.7   Reverberation spreading due to platform speed

9.12 Practical Reverberation Envelopes

By superimposing typical beam patterns onto the pulse spectrum, we can produce a set of reverberation envelopes for different platform speeds and target bearings (beams).

Figure 9.8 shows an envelope for a broadside beam and a platform speed of 10 knots. Note that, far from being noise limited, detections can take place against high levels of reverberation out to target dopplers of ±10 knots. The pulse lengths are necessarily long (for good noise-limited detection ranges and also for operation down to low magnitudes of doppler). The long pulse, however, results in a very low RI. Therefore, except against very low reverberation, detection performance can be poor until the target doppler exceeds platform speed (at broadside, but up to twice platform speed for the ahead and astern beams).

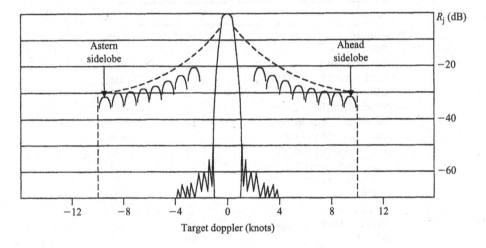

Figure 9.8 Typical reverberation envelope

Fortunately, this effect is only a problem when using *full-beam processing*. When half-beam processing is used, the phase differences obtained by comparisons between pairs of *codirectional half-beams* can be used to gate out the reverberation and target echoes in the sidelobes which, for a moving platform, would otherwise form a significant part of the detection background (i.e., it would be dominant over much of the likely target doppler spectrum).

Therefore, for *half-beam processing*, which will now be considered in detail, we effectively revert to backgrounds represented by the pulse spectrum unmodified by beam patterns, e.g., the Hamming pulse in Figure 9.5, although the main lobe spectrum will need to be broadened to allow for the effects of beamwidth, platform motion and the environment.

9.13   Full- and Half-Beam Processing

Full-beam and half-beam processing techniques are compared by postulating an active sonar using a simple omnidirectional, in azimuth, projector for transmission, and a 16-element (2 × 8 half-aperture) line array of hydrophones for reception, and having the parameters listed in Table 9.4. The sonar uses both LPFM and CW pulses. Identical pulse lengths and source levels are chosen so that, given matched filter processing, both pulse types have identical *noise-limited* detection performance.

Table 9.4   Sonar parameters

Parameter	Value	Units
Frequency	10	kHz
Pulse length, T	1	s
Source level, SL	210	dB
FM bandwidth	400	Hz
CW bandwidth	2	Hz
Beamwidth, full aperture	8	degrees
Beamwidth, half-aperture	16	degrees
Phase bin width (2 bins)	2	degrees
Reverberation index, RI		
FM, full beams	17	dB
FM, half-beams	23	dB
CW	−6	dB
DI (receive)	12	dB
Integrated target strength, TS	10	dB
$5 \log d$	10	dB
$5 \log n$	3	dB
Background noise level	50	dB

The performance of the two pulse types is then assessed for both full-beam and half-beam processing and the following benefits are shown to result from half-beam processing. The last two items are applicable to both FM and CW pulses.

- Enhanced FM detection performance for *all* doppler values

- Noise-limited CW detection performance once the doppler spectrum of the reverberation within the main lobe of a beam is exceeded

- Improved bearing accuracy

- Elimination of echo returns in adjacent beams: simpler autodetection and tracking processes and reduced operator workload

Improved bearing accuracy

Amplitude comparisons between full beams give, at best, resolutions or accuracies of around $1/4$ of the full aperture beamwidth, whereas phase binning can achieve $1/16$ of the half-aperture beamwidth, with a practical lower limit of $1°$. Autodetection and tracking processes clearly benefit from this improvement: fewer false associations are made, resulting in fewer false alarms; weapon guidance predictions are significantly enhanced; and weapon target acquisition rates are improved.

9.14   Beamforming

Segments of circular or cylindrical arrays are effectively reduced to equivalent linear or planar arrays by applying time delays to the elements, hence a short line of 16 elements spaced $\lambda/2$ is a useful and simple array with which to demonstrate the principles of full-beam and half-beam techniques.

Full beams are formed by the simple addition of all 16 elements, after shading and time delays, to form the required set of beams. (For a circular array the time delays are constant and different sets of elements are used to step around the array in azimuth.) The resultant full beamwidths will be of the order of $8°$ and spacing between beams is, in practice, somewhat less than this, to minimize scalloping losses. For simplicity, we will ignore scalloping loss and make the spacing equal to the full beamwidth ($8°$).

Pairs of codirectional half-beams are formed using the 2×8 elements of the line in a similar manner to the full-beam case. Both left and right half-beams now have widths of $16°$ and the spacing between each pair is also $16°$. The beam shapes are shown in Figure 9.9.

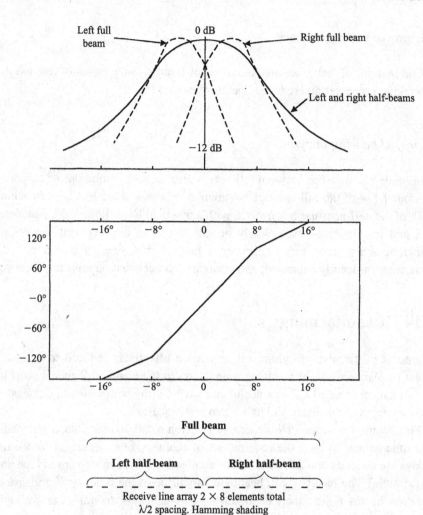

Receive line array 2 × 8 elements total
λ/2 spacing. Hamming shading
Full beamwidth 8° and all 16 elements used
Half-beamwidth 16° and 8 elements used

Figure 9.9   Half-beam and full beam: amplitude and phase plots

9.15   FM Phase Binning Process

Each pair of half-beams is processed as shown in Figure 9.10. The half-beam outputs are correlated against a replica of the transmitted pulse and the correlator outputs are bulk steered and added to form two full beams, together covering the same azimuth as the original codirectional half-beams (Figure 9.9). Phase comparison between correlator outputs is used to select the appropriate (left or right) full beam. The output of the selected full beam is then steered into one of 16 phase bins which together cover 16° of azimuth. (The left full beam output, if selected, will be steered into bins 1 to 8 and the right full beam into bins 9 to 16.)

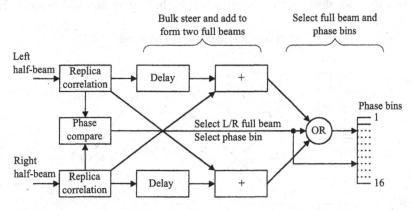

Figure 9.10   LPFM signal processing for one pair of codirectional half-beams

The target appearing in one phase bin, or overlapping into two phase bins (and here we will always assume two phase bins), is thus detected against a background – noise or reverberation – reduced by the ratio of the half-beam width to the width of two phase bins. The *reverberation index* is thus improved by 6 dB (8/2) compared with the full-beam output that would otherwise be used.

Integration of the independent samples (every $1/B$) from the phase binning process will give a *post-detection gain* until the integration time exceeds the target extent. If we assume the minimum target extent to be 15 m, we would sample and reset the phase bins every 20 ms and

$$\text{PD gain} = 5\log(BT_{\text{intgr}}) = 5\log(400 \times 20 \times 10^{-3}) = 4.5 \text{ dB}$$

Be careful with this PD gain when calculating the performance of a broadband system using the sonar equations. It is not appropriate if an *integrated* TS value is used in the sonar equation (the integration is then 'in the water'). The *peak* TS value, which is perhaps about 6 dB less, should be used. In view of the uncertainty surrounding TS measurements, although a PD gain certainly occurs, it is perhaps better to err on the side of caution and ignore it when making performance calculations.

9.16   CW Processing

Each pair of half-beams is processed as shown in Figure 9.11. The half-beams are frequency analysed to form N doppler channels. Left and right *full beams* are then formed for each doppler channel by a similar bulk steer and add process. Phase

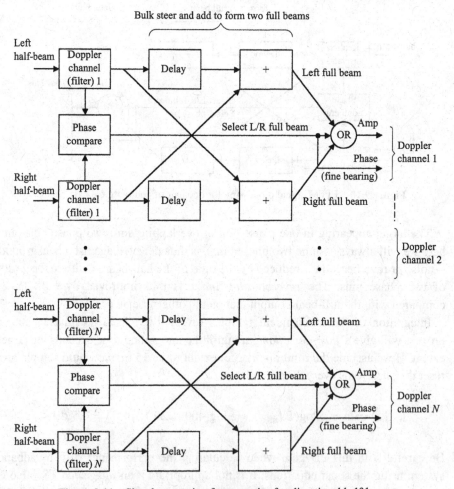

Figure 9.11   Signal processing for one pair of codirectional half-beams

comparison is again used to select the correct full-beam outputs from each doppler channel. These outputs are combined with the measured fine bearings (within a beam) and passed to data processing and display at a rate depending on the CW pulse length (e.g., every 250 ms for the 1 s pulse). There is no possibility of using phase bins for post-detection integration because the length of the CW pulse is already comparable with, or much greater than, the dimensions of the target. The long CW pulse has effectively integrated the echo energy in the water and therefore this time the *integrated* TS value should be used in the sonar equation.

However, there are still positive advantages in forming half-beams and determining target bearing by phase comparisons. The CW pulse will benefit from improved bearing accuracy and the elimination of echoes from adjacent beams in exactly the same manner as for the FM pulse. There is also an improvement in the CW detection performance, i.e., it becomes noise limited, immediately the doppler spectrum of the reverberation within the main lobe of a beam is exceeded, as explained in Section 9.12

9.17 Large Aperture Array

Figure 9.12 shows the beam pattern of one of a pair of codirectional half-beams formed using half the aperture of a 32λ array of 64 elements, together with the phase difference plot resulting from the complex cross-correlation of the pair of half-beams. Hamming shading is used. The $\pm 2°$ beamwidth of each half-beam corresponds to a phase difference of about $\pm 120°$. By using the bulk steer and phase binning process described earlier, phase differences greater than this can eliminate all returns from not only the sidelobes but also the skirts of the main lobe. Note that the phase difference remains greater than $\pm 120°$ for all values of azimuth beyond the 3 dB points of the main lobe. In practice, imperfections in the array, and unwanted curvature in the case of a towed array, may cause re-entrants to occur in the phase difference plot beyond the $\pm 120°$ limits which will result in returns through sidelobes. To minimize this, the sidelobe levels should be reduced by beam shading – Hamming shading in this example.

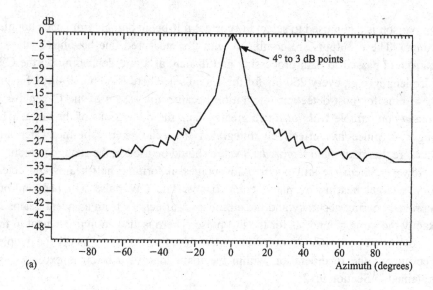

(a)

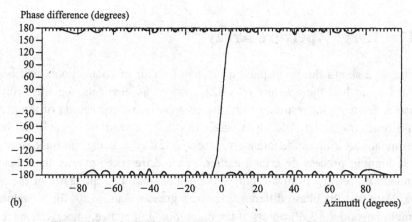

(b)

Figure 9.12   (a) Half-beam response and (b) phase difference plot

9.18 Detection Performance

Noise-limited performance

The noise-limited performance is identical for both pulse types and is given by

$$2PL = SL + TS - N + DI + 10 \log T - 5 \log d + 5 \log n$$

and using the parameter values of Table 9.4:

$$2PL = 210 + 10 - 50 + 12 + 0 - 10 + 3 = 175 \text{ dB}$$

(The integrated TS value is used for both pulse types and, for the FM pulse, this is justified by ignoring the PD gain due to the phase binning process.) If we assume spherical spreading and absorption, the range is 9 km.

Reverberation-limited performance of the LPFM pulse

The reverberation-limited performance of the LPFM transmission (or any broadband pulse) does not change with target doppler, including zero, and is given by

$$\boxed{10 \log R = RI - S_b - 41 + TS - 5 \log d + 5 \log n}$$

Using the parameter values from Table 9.4, the LPFM reverberation-limited ranges are given in Table 9.5 for three representative values of S_b.

Table 9.5 LPFM reverberation-limited range

	Reverberation-limited range (km)		
	$S_b = -30$	$S_b = -37$	$S_b = -40$
Full-beam processing	8*	40	80
Half-beam processing	32	160	320

*Note that, in high reverberation conditions, the noise-limited detection range, 9 km, is only achieved by using half-beam processing

- $S_b = -30$ dB, a high bottom S_b

- $S_b = -37$ dB, 75 per cent of UK inshore waters have S_b less than this value

- $S_b = -40$ dB, a low bottom, high surface value for S_b

At useful detection ranges (greater than say 5 km), submarines never subtend angles greater than 1°, hence any processing and display scheme based on *full beams* – which are inevitably at least a few degrees wide – will have no fine detail in the bearing dimension to help with detection and classification.

- *For noise and uniform reverberation*, where bearing extents are similar to the submarine's, detection is not improved by using half-beam processing. But the greater bearing accuracy and elimination of sidelobe returns will simplify and improve all subsequent processes.

- *For discrete reverberation*, where bearing and bearing extents can differ markedly from those of a submarine, detection is improved by using half-beam processing.

- *Classification and display*, against any background, are enhanced by the availability of fine bearing (phase bins) providing an extra dimension for echo shape recognition and improved bearing accuracy for tracking.

Reverberation-limited performance of the CW pulse

The reverberation-limited performance of the CW transmission varies with doppler and is given by

$$10 \log R = RI - S_b - 41 + TS - 5 \log d + 5 \log n - R_j$$

Note the additional term, R_j; this is the reverberation rejection and it determines how the performance of the CW pulse varies with doppler. (R_j is zero for FM pulses.) We are effectively calculating the detection range of a target with zero doppler and then multiplying it (adding dB) by some factor determined by R_j.

For the CW transmissions, we are less interested in the reverberation-limited range and pose the question, How much doppler before the detection is noise limited? To answer this, assume the reverberation-limited range is 20 km (i.e., always longer than the noise-limited range) giving $10 \log R = 13$ dB. Then solve the reverberation-limited equation for R_j, again using the parameters of Table 9.4. The target dopplers required to achieve the R_j values (Table 9.6) are then obtained from Figures 9.6 and 9.8, but adding another 0.7 knot as suggested in Section 9.5.

Note that the doppler magnitudes required using full-beams are for a moderate platform speed, 10 knots and assume quite low sidelobes (20 to 30 dB down). In practice, noise-limited detections will frequently only be achieved for target

Table 9.6 Target doppler

S_b (dB)	R_j (dB)	Doppler (knots)
Full-beam processing		
−30	−27	8
−37	−20	3
−40	−17	2
Half-beam processsing		
−30	−27	1.1
−37	−20	1.0
−40	−17	1.0

dopplers which exceed the platform speed for targets on the beam, and twice platform speed for targets ahead or astern.

Half-beam processing is superior to full-beam processing of CW pulses. Detections remain noise limited down to target dopplers of about 1 knot for a 1 s pulse. There is little to gain – in performance against low dopplers – from using longer pulses. This is because the factors other than pulse length which cause reverberation spreading – finite beamwidth, platform motion and the environment – then become dominant.

Because of the target doppler requirement, it is operationally unsafe to use only a CW pulse for any active sonar – unless there is a high probability of sufficient target doppler – and the preferred mode of operation is to transmit both FM and CW pulses every ping.

9.19 Noise- and Reverberation-Limited Detection Ranges

Figure 9.13 is based on the parameter values of the hypothetical 10 kHz sonar. Propagation is assumed to be limited by spherical spreading and absorption. Both noise- and reverberation-limited detection ranges are plotted against TS for the LPFM pulse. Note the significant difference in slopes. This is because the reverberation-limited range is a function of $10 \log R$, whereas the noise-limited range is a function of $40 \log R$ (i.e., R^4 compared with R).

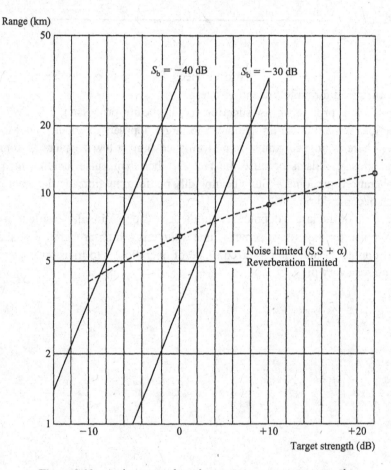

Figure 9.13   Active sonar: detection ranges versus target strength

Noise-limited range is therefore comparatively insensitive to changes in TS or N (a change of 3 dB results in about 20 per cent change in range), whereas when reverberation limited a similar change in TS or S_b will change the range by 100 per cent. The comparison indicates the importance of attempting to ensure that the sonar is always noise limited. Clearly, there will be constraints which will often make this impossible, but it must be the underlying aim in the design of any active system.

9.20   Ambiguity Diagrams

The concept of ambiguity diagrams provides further insight into the properties
of the pulses used in active sonar. Active sonars measure the range (time) and
doppler (frequency) components of a target by cross-correlating overlapping
segments of the received signal with a set of stored references. Each reference is a
replica of the transmitted pulse individually modified in time and/or frequency.
Sufficient references are used to cover the expected extent of target doppler.

Ambiguity functions provide a means of comparing the range and doppler
resolutions achievable from different pulse types and of determining the references
required for a given sonar task. Figure 9.14 shows the output power from a correlator
versus the time and frequency displacements of the received signal. The half-power
(3 dB down) extent of the correlator output is given approximately as follows:

- In time (t) it is $1/B$

- In frequency (δf_0) it is $1/T$

Note that the output power is negligible at twice these displacements.

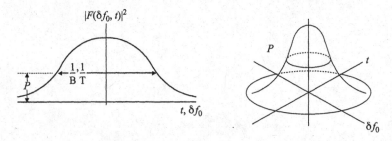

Figure 9.14   The ambiguity function

The 3D representation shows the correlator output above a 2D surface (t, δf_0).
The intersection at the detection threshold P (half-power) defines an ambiguity
contour within which a target cannot be located unambiguously, since all (t, δf_0)
combinations result in detections.

A plot of this contour versus δf_0 and t is known as the ambiguity diagram for
the pulse waveform. The ambiguity diagram, therefore, indicates the accuracies
and resolutions in range and doppler which are achievable from a given pulse.

The treatment which follows has been stripped of the somewhat difficult
mathematics necessary for even approximate derivations of the ambiguity diagrams for CW, FM and PRN pulses and concentrates on results and practical
examples.

CW pulses

By equating the ambiguity function for a rectangular CW pulse to 0.5 (half-power) and first putting $t = 0$ and then $\delta f_0 = 0$, we obtain $\delta f_0 = 0.88/T$ and $t = 0.6T$, and the resulting ambiguity diagrams for long and short CW pulses are sketched in Figure 9.15. The area of the elliptical ambiguity diagram – for any length of pulse – is approximately $0.6T(0.88/T)(\pi/4) = 0.4$. The values for δf_0 and t can be used to determine the resolution (extent of the ambiguity) of the pulse and the number of references (in this case matched filters) required to cover the expected target doppler.

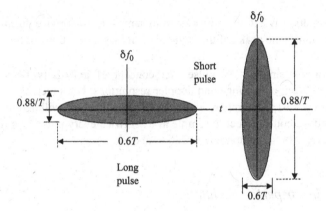

Figure 9.15   Long and short CW pulses

Example 9.4
A sonar transmits a 4 kHz CW pulse of duration 0.5 s and is required to detect targets with dopplers up to ± 20 knots. What are the range and doppler resolutions of the pulse and how many references (filters or FFT cells) are required?

We have

$$\delta f_0 = \frac{2vf_0}{c} = \frac{0.88}{T}$$

Therefore

$$v = \frac{0.88c}{2Tf_0} = \frac{0.88 \times 3000}{2 \times 0.5 \times 4000} = 0.7 \text{ knot}$$

Note that here c must also be in knots, 3000. The doppler resolution (ambiguity) of the pulse is 0.7 knot and therefore to cover ± 20 knots requires $40/0.7 = 57$ references.

We have

$$t = 0.6T = \frac{2R}{c}$$

Therefore

$$R = \frac{0.6 \times 0.5 \times 1500}{2} = 226 \text{ m}$$

The range resolution is 226 m and we need to sample at least twice during this range to ensure we capture the peaks of the signal, i.e., at least every $0.6 \times 0.5 = 300$ ms.

Note that, for the simple CW pulse, the concept of ambiguity functions is not necessary to determine the range and doppler resolutions. For a rectangular CW pulse the 3 dB width is $0.91/T$ (compare with $0.88/T$ above), which will determine doppler resolution, and we could expect to sample at least twice every pulse length, i.e., every 250 ms, or every 188 m (compare with 226 m above).

FM pulses: linear period modulation

In linear period modulation (LPM) doppler changes the received signal in frequency only, causing an overlap loss with the reference, and a slope mismatch does not occur whatever the magnitude of BT. By equating the ambiguity function for the LPM pulse to 0.5 (half-power) and putting first $t = 0$ and then $\delta f_0 = 0$, we obtain

$$\delta f_0 = 0.88/T \quad \text{and} \quad t = 0.88/B$$

The endpoints of the ambiguity diagram (Figure 9.16) are $\delta f_0 = \pm 0.3B$ and $t = \pm 0.3T$. And for any length of pulse, the area of the ellipse is approximately $0.6B(0.88/B)(\pi/4) = 0.4$.

The doppler resolution (ambiguity) of the pulse is given by the maximum δf_0 value of the diagram:

$$\delta f_0 = \frac{2vf_0}{c} = 0.6B \quad \text{therefore} \quad v = \frac{900B}{f_0} \quad \text{(knots)}$$

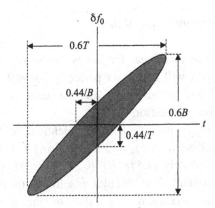

Figure 9.16 Ambiguity diagram: LPM pulse

The range resolution of the pulse is given by the intersections of the contour with the time axis:

$$t = \frac{0.88}{B} = \frac{2R}{c} \quad \text{therefore} \quad R = \frac{660}{B} \quad \text{(metres)}$$

Example 9.5

A sonar transmits a 10 kHz LPM pulse of 500 Hz bandwidth and 1 s duration. What are the range and doppler resolutions of the pulse? What is the maximum target doppler for 3 dB correlation loss?

$$R = \frac{660}{B} = 1.32 \text{ m} \qquad v = \frac{900B}{f_0} = 45 \text{ knots}$$

Because the doppler resolution is derived from the 3 dB contour, then the maximum target doppler for 3 dB loss is ± 22.5 knots (half of the resolution). The small correlation losses resulting from large target dopplers explain why the LPM pulse is known as a doppler-invariant pulse.

Once again a simpler derivation is possible. The range resolution is simply the inverse of the bandwidth, i.e., 2 ms or 1.5 m. There will be a 3 dB correlation loss when only 0.707 of the signal overlaps the reference, which is equivalent to a doppler shift of ± 147 Hz or $\pm 147/6.9 = \pm 21$ knots (from $\Delta f = \pm 0.69$ Hz per knot per kHz), which is in good agreement with the previous result of ± 22.5 knots.

FM pulses: linear frequency modulation

In contrast to an LPM pulse, a linear frequency modulated (LFM) pulse which is changed by target doppler will no longer perfectly correlate with any part of the zero doppler reference and its ambiguity diagram computation assumes that the slope mismatch is the limiting factor.

We have $\delta f_0 = 0.88/T$ and $t = 0.88/B$, identical to the LPM case. The endpoints of the contour are now $\delta f_0 = 3.5 f_0/BT$ and $t = 3.5 f_0/B^2$ and the area of the ellipse is approximately $(3.5 f_0/BT)(0.88/B)(\pi/4) = 2.4 f_0/TB^2$, which is no longer a simple constant. The doppler resolution (ambiguity) of the pulse is again given by the maximum δf_0 value of the diagram (Figure 9.17), $\delta f_0 = 2 v f_0/c = 3.5 f_0/BT$, therefore $v = 5200/BT$ knots. The range resolution is $R = 660/B$ metres, as before.

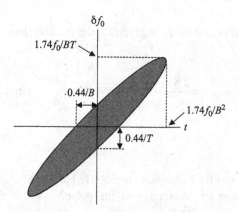

Figure 9.17   Ambiguity diagram: LFM pulse

Note that, because of the assumption that slope mismatch is the limiting factor, the equation for doppler resolution only gives valid results for large BT products. For small BT products, a better result is obtained from the simpler derivation used for the LPM pulse: calculating a doppler shift equivalent to 0.29 times the signal bandwidth.

Example 9.6

A sonar transmits a 10 kHz LFM pulse of 500 Hz bandwidth and 1 s duration. What are the range and doppler resolutions of the pulse? How many references are required to limit correlation loss to 3 dB for target dopplers up to ±30 knots?

We have

$$R = 660/B = +0.66 \text{ m}$$

$$v = 5200/BT = 5200/500 = 10.4 \text{ knots}$$

Therefore we require $60/10.4 = 6$ references

The area of the contour is $2.4 f_0/TB^2 = 0.1$, which is smaller than the corresponding area of 0.4 for the LPM pulse. The smaller the area, the smaller the ambiguity of the pulse (the better the resolution).

As we have seen, the smaller area of the contour for the LFM pulse indicates an improved doppler resolution – about 10 knots instead of 45 knots for the LPM pulse. The smaller ambiguity of the LFM pulse explains why it is sometimes known as a doppler-intolerant pulse. Note, however, that while it may be intolerant from the viewpoint of limiting correlation loss, the pulse only provides a poor measure of the target doppler. (Compare this with a CW pulse of the same duration and frequency, which would have a doppler resolution of 0.14 knot.)

Example 9.7
A sonar transmits 10 kHz LPM and LFM pulses of bandwidth 100 Hz and pulse duration 1 s. What are the doppler resolutions of the pulses?

Suppose we begin like this:

- LPM $v = 900 B/f_0 = 9$ knots
- LFM $v = 5200/BT = 52$ knots

This is clearly an invalid result; the LFM pulse is apparently more tolerant of doppler than the LPM pulse of identical bandwidth and duration. A better result is now obtained from the equation used for the LPM pulse. Both pulses, then, have the same doppler resolution of 9 knots.

Example 9.7 highlights the problem of how to decide which equation to use. For practical values of B and T, a valid result will be obtained using

- $v = 5200/BT$ knots for $BT \geqslant 500$
- $v = 900 B/f_0$ knots for BT $\leqslant 100$

For intermediate values of BT, or indeed for any BT value, a recommended, robust approach is to use both equations and take the result of smaller magnitude.

PRN pulses

An alternative to a broadband FM pulse is the pseudorandom noise (PRN) pulse. Its bandwidth will be close to or identical to the bandwidth of an FM pulse designed for a similar task. The frequency–time structure of a PRN pulse will change in a random but known manner from ping to ping. Hence replica correlation will still be possible.

By equating the ambiguity function for the PRN pulse to 0.5 and putting $t = 0$, $\delta f_0 = 0$, we obtain the 3 dB (half-power) ambiguity diagram (Figure 9.18), where $\delta f_0 = \pm 0.44/T$, which is identical to the expression for a CW pulse and $t = \pm 0.44/B$, which is identical to the expression for FM pulses.

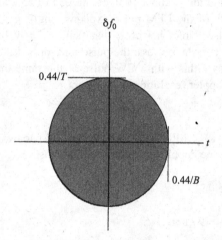

Figure 9.18   Ambiguity diagram: PRN pulse

The area of the circle is approximately $(0.88/T)(0.88/B)(\pi/4) = 0.6/BT$. The doppler resolution of the pulse is given by $\delta f_0 = 2vf_0/c = 0.88/T$ or $v = 0.88c/2Tf_0$ (as for CW pulses). The range resolution is given by $t = 0.88/B$ or $R = 660/B$ metres (as for FM pulses).

The PRN pulse therefore combines the desirable properties of both CW and FM pulses, i.e., *simultaneous good doppler and range resolutions*. If BT is less than 2400 divided by the maximum target doppler, the references may simply be frequency shifted, otherwise time-compressed or time-expanded references must be used.

Compared with using a combination of CW and FM pulses, the drawbacks of PRN are the need for a large number of references, each of which must be sampled at least twice in a period equal to the inverse of the bandwidth (at least every 1 ms for a bandwidth of 500 Hz), and possible complications or compromises necessary in the display of the data.

The processing power requirement is unlikely to be a problem with current technology but, as always, availability of an adequate bank of PRN trials data is a strong argument for retaining combined CW and FM sonars.

Example 9.8

A sonar transmits a 5 kHz PRN pulse of 400 Hz bandwidth and 1 s duration. What are the range and doppler resolutions of the pulse? How often must the signal be sampled and how many references are needed for a maximum target doppler of ±20 knots? We have

$$v = \frac{0.88c}{2Tf_0} = \frac{0.88 \times 3000}{2 \times 5000} = 0.26 \text{ knot}$$

To cover ±20 knots, we need $40/0.26 = 154$ references, each spaced 0.26 knot and 0.26 knot wide.

$$R = 660/B = 1.65 \text{ m}$$

and $BT = 400$, which is greater than $2400/20$, so time-compressed or time-expanded references must be used.

Statistically independent signal samples are separated by the inverse of the bandwidth. (Note that the approximate solution using the ambiguity function is $0.88/B$, very close to the inverse of the bandwidth.) Here the inverse of the bandwidth is 2.5 ms, and to ensure that peaks of the signal are not missed, sampling should occur at least twice during this period, say, every 1 ms.

$$\text{Area of the contour} = 0.6/BT = 0.6/400 = 0.0015$$

This area is very much smaller than for similar duration CW or FM pulses and confirms the low ambiguity (good resolution) in both range and doppler of this class of pulse.

Table 9.7   Range and doppler resolution for different pulse types

Pulse type	Range resolution (m)	Doppler resolution (knots)
CW	$450T$	$1350/Tf_0$
LPM	$750/B$	$900B/f_0$
LFM ($BT < 100$)	$750/B$	$900B/f_0$
LFM ($BT > 500$)	$750/B$	$5200/BT$
PRN	$750/B$	$1320/Tf_0$

Range and doppler resolutions of CW, FM and PRN pulses

Table 9.7 collates expressions for the range and doppler resolutions for the pulse types considered. For intermediate BT values of LFM pulses, and preferably for any values, calculate the doppler resolution using the two alternative formulae, as discussed earlier, and use the smaller of the two results.

9.21   Very Long Pulses

A very long pulse will, in practice, have a processing gain somewhat less than its theoretical value. For pulses between 1 and 10 s duration, the degradation may be of the order of 3–6 dB and is the result of instabilities in the environment, the platform and the target. For example, a platform course change which resulted in the target doppler changing during reception of an echo from a CW pulse, would spread the echo energy into adjacent matched filters. Pulses up to, say, 2 s duration may, at least under favourable conditions, have acceptable correlation losses but, in order to maximize the transmitted energy and thus improve noise-limited performance, the alternative of transmitting a train of pulses should be considered.

Example 9.9

To improve noise-limited performance, the term $10 \log T$ in the active, noise-limited sonar equation must be increased. Suppose a 10 s pulse is transmitted, then $10 \log T = 10$ dB. But if we assume the correlation loss to be 6 dB, the effective gain over a 1 s pulse is only 4 dB. If the long pulse is replaced by 5 pulses, each of 2 s and separately processed, then $10 \log T = 3$ dB but this can be followed by post-detection integration, giving a further $5 \log 5 = 3.5$ dB gain. The effective gain over a 1 s pulse is now 6.5 dB, an improvement of 2.5 dB over transmitting the single long pulse.

9.22   Operational Degradation Factor

A figure of 4 dB is often suggested for the degradation in operational performance of a modern computer-assisted sonar, compared with its theoretical performance. However, 4 dB typically equates to a range reduced to about 0.7 of theoretical range in noise-limited conditions, but only 0.4 of theoretical range in reverberation-limited conditions.

The factors contributing to the DF are unlikely to be identical for both reverberation-limited and noise-limited conditions and, particularly in highly reverberant shallow waters, it is the extra time required for the elimination of false alarms and final classification which may decide the initial detection range (the range may just as likely remain constant, or even increase, depending on the relative velocity and bearing of platform and target during this time).

Fluctuations in TS and PL from ping to ping are frequently very large, and marginal initial detections seldom, if ever, occur. A contact is typically non-existent during one ping and very strong on the next ping. Because of such uncertainties, the whole concept of degradation is debatable for active sonars; but if it must be applied, the recommended approach is to calculate both noise-limited and reverberation-limited detection ranges and then to reduce the smaller by a factor of 0.7 to allow for a final classification.

9.23   Active Displays

Virtually all active sonar displays conform to the general format of Figure 9.19. Range (time from transmission) is displayed on the vertical axis and bearing, or bearing + doppler, is displayed on the horizontal axis. Most active displays will have 'raw sonar data', i.e., post-detection samples, forming a background to 'synthetic sonar data', i.e., data on contacts generated by automatic detection and classification processes.

In Figure 9.19 contact 3 is a possible submarine. The vertical stroke of the marker indicates the expected position of the contact during the next ping, and the magnitude of the horizontal stroke indicates the likelihood that the contact is genuine. Computer-generated data on the most likely contacts (10, 20, 40?) are displayed in the form of totes – alphanumerics of range, bearing, course, speed, classification, confidence, etc., on all likely contacts, or on the history of a selected contact. Typical active displays are best illustrated by postulating a hypothetical but realistic sonar with the parameters in Table 9.8.

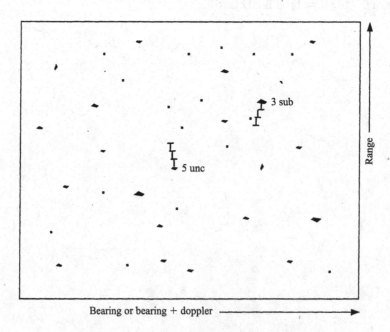

Figure 9.19   Active displays: general format

Table 9.8 Parameters of a typical sonar

Parameter	Value	Units
Frequency	10	kHz
Pulse length, T	300	ms
FM bandwidth	400	Hz
CW bandwidth	7	Hz
Beamwidth, half-aperture	16	degrees
Beam spacing	11.25	degrees
No. of beams, 360° cover	32	
Doppler channel width	1	knots
Doppler cover	±30	knots
FM increments		
classification (3 ms)	2	m
surveillance (20 ms)	15	m
CW increment (133 ms)	100	m
Phase bin width	1	degrees
Range scales	4, 8, 16	km

FM surveillance display

This typical sonar uses half-beam processing. The 1 ms correlator samples are placed in phase bins and integrated to 3 ms for classification (3 ms is the inverse of the bandwidth and the minimum time for independent samples). The 1 ms samples are integrated to 20 ms for surveillance, i.e., roughly matching the minimum target dimension. The FM surveillance display must therefore display the integrated contents of the phase bins for each beam as a function of range (time).

RANGE
512 pixels

BEARING
512 pixels
32×16 phase bins

Range scale	4	8	16
Range increments	267	533	1067
Pixels per increment	2	1	0.5

For the 4 km range scale, display the same information in two consecutive lines. For the 16 km range scale, use one line to display two consecutive range increments. (The maximum ranges displayed will be slightly less than the nominal range scales.)

CW surveillance display

The CW processing provides independent samples every 133 ms. This is equivalent to a 100 m range increment, which is already large compared to the minimum target dimension, hence it requires no integration before display.

RANGE
512 pixels
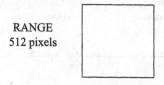

BEARING, DOPPLER
512 pixels

Range scale	4	8	16
Range increments	40	80	160
Pixels per increment	12	6	3

Note that, because of the small bandwidths associated with the CW pulses used for submarine detection, range resolution is poor and, for all range scales, the same information is displayed in several lines if the display is to be the same size as the FM display. Alternatively, these surplus lines could profitably be used to display ping history. Whatever solution is adopted, equal height displays are clearly desirable to allow the operator to readily associate FM and CW echoes from the same target.

The maximum number of increments in the horizontal dimension (bearing + doppler) is $32 \times 60 = 1920$. This number must be reduced by ORing beams and/ or doppler channels to match the available number of pixels, e.g., 16 beams $\times$ (30 doppler channels + 2 spaces) = 512. The doppler channels may be varied in width, increasing with the magnitude of the target doppler, to reflect both the improved performance and the lower probability of high magnitude target dopplers.

The range, bearing and doppler samples, however, will be passed to the automatic detection and classification processes at their full resolutions.

History displays

Active history displays are analogous to the *BT* and LOFAR passive displays, where successive time samples are replaced by successive ping samples. Figure 9.20 shows the simplest form of history display, used by the early electromechanical range recorders. A slope equivalent to the platform relative velocity will indicate reverberation or a zero doppler contact. Other slopes indicate non-zero doppler contacts. Noise will seldom correlate from ping to ping. Derivatives of this display are still used successfully in modern sonars and are not lightly disregarded.

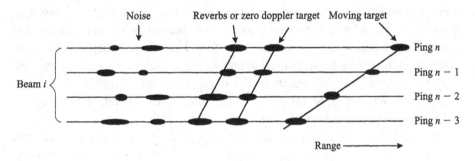

Figure 9.20 Range recorder trace

Geographically stabilized displays

Geographically stabilized displays are a modern form of history display, made possible by the processing power now available. If all the pixels of ping $n - 1$ are corrected for own ship movement in range and bearing and then summed, at the display, with the corresponding pixels from ping n, the display will be geographically stabilized. Both reverberation and stationary targets (truly stationary, not simply having zero relative velocity) will be displayed at the same coordinates for all pings, but moving targets will change position from ping to ping.

Stationary targets will increase in brightness, but so will reverberation and therefore, at least in reverberation-limited conditions, performance against stationary targets will be little enhanced. Moving targets will produce tracks on the display which will be easier to detect and classify. To avoid display whiteout after several several pings, the individual pixel amplitude (brightness) values of the samples from earlier pings must be decreased with time.

9.24   Unified Detection and Classification

The traditional approach to the design of an active sonar system treats the detection and classification functions as separable and consecutive, but in reality they are interdependent. Design decisions are then based on detection alone, i.e., simply finding signals in noise or reverberation based on amplitude only, and this may not be optimum for the composite detection and classification task.

The traditional approach – optimizing detection before considering classification – can be effective against noise and low reverberation but less so against high discrete reverberation (clutter) such as is often present in coastal waters. This discrete reverberation is the source of many false alarms, and a major problem not properly addressed by the traditional approach where the sonar equations are used to define the probability of detection (P_d) at a specified probability of false alarms (P_{fa}) based only on S/N ratios. Classification, which should be seen as the reduction of false alarms, comes later.

The unified approach recognizes that detection and classification are not separable. Design decisions must recognize this and the total task must be optimized, not initial detection at the expense of classification. The parameters fundamental to this unified approach are *bandwidth* and *beamwidth*.

9.25   Bandwidth

For detection we need to match the minimum target dimension (d). Suppose for a submarine this is 10 m, then the resolvable pulse length is $t = 13$ ms from $d = ct/2$; as we have already seen, t can be much less than the actual pulse length. Therefore bandwidth $= 1/t = \sim 75$ Hz. For classification we need to resolve the structures of the target and any non-targets (false alarms). This will require a higher resolution, to say 1 m, giving bandwidth $= 750$ Hz.

If the target is small, e.g., a mine, the required resolution will be much higher, say 0.01 m, giving bandwidth $= 75$ kHz. We could use either a CW pulse of duration $1/B = 13\mu$ s or a longer, perhaps 10 ms, broadband (FM type) pulse.

The point to emphasize here is that bandwidth is decided by the dimensions and structure of the target and the false alarms. *There is nothing to be gained by increasing the bandwidth further.* Note that if there are reflecting objects much smaller than the target, they will be irrelevant because their target strengths will be too small to compete with those of the target and comparable false alarms.

9.26 Beamwidth

Just as bandwidth determines the resolution in the range dimension, so does beamwidth in the bearing dimension. Again, for detection we would like to match the minimum target dimension, which is now the minimum angle subtended at the array by the target. This is directly proportional to range, and for a submarine at a range of at least 5 km this will always be less than about 1°. Given practical arrays, the smallest beamwidth will be at best about 4° and matching will only occur at around 1–2 km, depending on submarine aspect. (Smaller beamwidths are possible using towed arrays, but only at bearings within, say, normal ±60°.)

Figure 9.21(a) shows, in B scan format, submarine targets at two different ranges and a patch of discrete reverberation which might, for example, represent a rocky outcrop in shallow water. Only one beam of the display is shown: the beamwidth is 4° and the range is from 0 to 5 km. Note that the width of the display is proportional to range: at 5 km the width is about 270 m and therefore the submarine at this range has a bearing spread of about 1°, but at 1.7 km the width is 90 m and its bearing spread is much greater, about 3°.

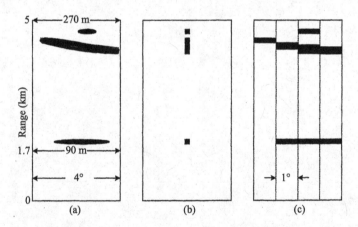

Figure 9.21 Comparison of bearings and bearing spreads for submarines and false alarms

Figure 9.21(b) shows one beam of a sonar display in the same B scan format and produced using full-beam processing. Because full-beam processing cannot resolve bearing within a beam, the targets and the false alarm will all be given the same bearing and no values at all for bearing spread. The submarine targets and the reverberation patch, therefore, all appear at the same bearing and with the same (unknown) spread in bearing. Note, however, that the differences in range spread are displayed.

Figure 9.21(c) shows one beam of a sonar display again in the same format but produced using half-beam processing. We can now resolve bearing within a beam, and the submarine targets and the reverberation patch are all displayed with their correct bearings and bearing spreads. We now have, within the sonar's limits of resolution, a fair approximation to the overall shape and position of contacts and false alarms.

At 5 km range, the submarine is in one phase bin whereas the reverberation occupies four phase bins. Since, at this range, a submarine cannot spread into more than two bins, the bearing spread is a useful discriminant between targets and false alarms. At closer ranges, this discriminant may not be so useful. The target spread, however, when considered together with other parameters such as target motion, will still be a strong classification feature. For classification we again need to resolve the structures of the target and any non-targets (false alarms) in the bearing dimension. Discrete reverberation (clutter) can have much greater bearing spreads than submarines, and in order to exploit this difference for classification (false alarm reduction), we require a higher bearing resolution up to, say, 1/10 of the beamwidth. A practical minimum value for the composite detection and classification task might be 0.5°.

Full beams, using the greatest possible aperture of the array, cannot achieve such small beamwidths, therefore detection and classification processing using signals from full beams is not optimal against a background of high discrete reverberation (clutter).

However, against a background of noise or uniform reverberation and only if the other important advantages of the alternative, half-beam processing, are ignored, then full-beam processing, which uses the highest possible array DI, is marginally better: the 3 dB greater DI provides an extra 1.5 dB of allowable propagation loss to use in the sonar equations.

In summary, half-beam processing can be used to effectively achieve small beamwidths and therefore detection and classification processing using signals from half-beams is optimal against a background of high discrete reverberation (clutter). The effective beamwidth is the size of a phase or bearing bin.

The phase bins, as recipients of the output signals from the process, resolve the structures of both targets and false alarms in both range and bearing dimensions, to provide the best possible data for the composite detection and classification task.

9.27 CADAC

The output from the signal detection process is a time series of samples from every beam of the sonar, and for modern high resolution sonars there are many samples every ping. Suppose $P_d = 0.5$ and $P_{fa} = 10^{-5}$ (this is for the simple signal detection process only). For the typical sonar which was used to illustrate active displays, the FM display has $512 \times 500 = 2.5 \times 10^5$ pixels (samples). Therefore there are 2.5 false alarms every ping, or 15 false alarms per minute! What appears to be a very low P_{fa} has become an intolerably high rate of false alarms. The post signal detection processes of CADAC (computer-aided detection and classification) and display will help the operator to reduce this to an acceptable false alarm rate to pass to the command system.

9.28 Levels of CADAC

The post signal processing functions (CADAC and display) host many processes, some of which have traditionally been thought of as detection and others as classification. All of these processes contribute to a final classification and no individual process is purely for detection or classification. (Compare with the argument above that, for signal processing, detection and classification should not be considered separately.) The combined CADAC and display task may be split into several levels, each level building on the earlier ones. It would be wrong to say that any level detects or any level classifies, rather that every level contributes to the final classification.

Level 1

Level 1 is the lowest level; it detects signals within single cells; the signals may originate from vessels, noise or reverberation. A fixed threshold is normally used, at around 10–13 dB above the average of the background (the threshold may be lowered to dig deeper into the background in areas of interest, or increased to inhibit inputs from improbable areas).

Level 2

If the sonar resolution is finer than the typical size of a contact (including false alarms such as discrete reverberation) then returns will occur in several adjacent range and bearing cells. Clustering these returns together forms the second level of classification, since some estimate of the dimensions and shape of the contact can be formed from the distribution of returned energy in the contiguous cells. This estimate is commonly known as the *individual weight of evidence* of a contact. The parameters used to make this estimate include cell amplitudes, total energy, range, range spread and bearing spread. At level 2, additional processes can be applied to the cluster of individual returns making up a contact. There are, broadly speaking, two ways of doing this: statistical analysis and amplitude profiles (Sections 9.30 and 9.31).

Level 3

Level 3 examines how well contacts associate from one ping to another by seeking consistent tracks. Goodness of track and individual weights of evidence of the contacts forming the track help to build up a parameter known as the *weight of evidence* (WoE) of a contact. Courses and speeds of all contacts are also calculated at this level (the speed of a contact is clearly an important classification clue). Target doppler, if available, should be input to the WoE at this level.

Level 4

At level 4, information gathered from all other possible sources is used to assist in the final classification of all contacts. Here are some possible sources:

- Other modes of the sonar such as passive or intercept
- Other sonars on the same platform
- Sonars and other sensors on consort vessels or aircraft
- Sonobuoys
- Radar

Many of these sources will not help with the classification of the submarine itself. They will, however, materially assist in the reduction of false alarms. The reduction of false alarms can itself be viewed as classification. If all false alarms are eliminated, surviving contacts must be submarines.

Level 5

Level 5, the final level, includes all the experience, knowledge of tactics, strategy, and intelligence data possessed by the operator and the command, and it is only at this, human, level that the classification can be positively confirmed. For this reason, any automated classification system, no matter how clever its algorithms, should be thought of as a classification aid not an automatic classifier.

9.29   CADAC and Pulse Features

For long CW pulses, the range resolution cell is very large – hundreds of metres – and clearly provides no useful discrimination at CADAC level 2. A CW detection, however, is highly unlikely without some target doppler to ensure that target echoes fall into regions of the reverberation spectrum where the reverberation energy is low. The important classification clue of target doppler is therefore always available simultaneously with CW detections.

For the FM pulses, the range resolution cell is small compared with the typical extent of a contact and therefore it is possible to make improved estimates of the dimensions and echo profile of the contact. Furthermore, half-beam processing, by improving bearing resolution, provides additional information (bearing spread) towards this estimate, which can help to discriminate between targets and false alarms at CADAC levels 1 and 2.

For both pulse types, half-beam processing improves bearing accuracy and eliminates echo returns in adjacent beams, from which ping-to-ping association and tracking processes will benefit (CADAC level 3).

9.30   Statistical Analysis

Statistical analysis examines the joint distributions of amplitudes, ranges and bearings of the resolved echoes of a large number of previously identified contacts. Using specialized pattern analysis and recognition software, suitable discriminants can be found to partition the returns into the desired classes. For example, it might well be that the distribution of amplitudes with range for a submarine contact follows a positively skewed probability function whereas a surface ship gives a Gaussian distribution. The preliminary analysis phase examines all the possible statistical measures and identifies the minimum subset which provides the required discrimination among the known contacts. As long as the statistics of contact echoes when the system enters service remain similar to those of the preliminary training set, then it will be able to make successful attempts at classification into the original categories.

9.31   Amplitude Profiles

The amplitude profiles approach imitates the way an experienced operator assesses the contact echo. If the resolution is sufficiently fine (1 m or so) then the echo from a contact will display some detail of the reflecting structure of the contact. The amplitude of the returns along the length of the contact will be proportional to the reflectivity of the contact's structure. A submarine may well return strong echoes from the bow and the fin, and weaker echoes from the hull and the casing. These reflections yield a set of strongly aspect-dependent amplitude/range profiles which, after normalizing and 'clean-up' processes, are compared with memorized profiles of known contacts to give an indication of the contact's type.

The method has the advantage that it can 'learn' new profiles from just a single exposure to a particular contact, and can dynamically adjust its repertoire of stored profiles in accordance with the prevailing conditions. Furthermore, it has the important attribute of being able to indicate to the operator why it has made its hypothetical choice between alternative classifications in terms that the operator can understand, rather than as obscure statistical technicalities.

There is a case for providing both statistical analysis and amplitude profile aids to classification. The statistical analysis classifier is probably superior in classifying those targets with which it has been trained. But it is not robust to changes in processing, it is sensitive to the environment (e.g., using different algorithms for deep and shallow water), and from an operator viewpoint it makes its decisions 'in the dark'.

The amplitude profile classifier is perhaps somewhat less than optimum when operating against the specific targets with which the statistical analysis classifier has been trained. On the other hand, it is robust to changes in signal processing, it is insensitive to the environment, it can rapidly 'learn' new profiles from just a single exposure to a new target, and because the profiles are displayed to the operator, it makes its classification in terms that the operator can see and understand.

9.32   Multipath Affects Classification

To interpret any classification, it is important to understand contact clusters and to ask: What are the effects of multipath propagation on the individual echoes returning from a target? Multipath returns are unlikely to significantly increase the range or bearing extents of a cluster; path differences for similar strength returns are unlikely to be greater than a few milliseconds. They can, however, confuse the structure of the amplitude profile of a contact, by smearing and moving the highlights, or even resulting in new ones, e.g., glints from features of the target which are at normal incidence to the additional paths.

The operation of either type of classifier can therefore be impaired by multipath. It has been suggested that the effects of multipath propagation could be deconvoluted ahead of the classification procedures which are affected by them. This relies on a more precise knowledge of propagation than is likely in practice. Perhaps a more promising approach would be to attempt to eliminate or reduce the confusing returns. Narrow, steered, vertical beams could be employed, or if the vertical beamwidth cannot be less than, say, 10° then phase binning in the vertical dimension could be tried.

9.33   Simple Multipath Example

Classification displays often plot the amplitudes of individual returns (echoes) from a target against range (target extent), to produce a range/amplitude profile of the target. Multipath returns distort this profile in two ways (Figure 9.22):

- Point *a* on the target submarine returns signals to the source via two paths, *A* and *AA*, for this simplified case (Figure 9.22). The same point therefore puts returns in two separated cells along the profile. Signals along paths *AA* and *B* arrive simultaneously but from different points (*a* and *b*) on the target, and are detected against a background of reverberation returning along path *C* and also arriving at the same time. Different point sources therefore put returns in the same cell of the profile.

- An echo from point *b* returned via the bottom will extend the length of the profile. The magnitude of this extension will increase with range – as path differences and the number of reflections from the boundaries increase.

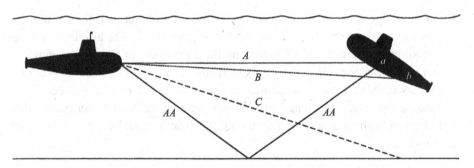

Figure 9.22   Multipath example

Multipath propagation increases the total energy returning to the source from the target without increasing the reverberation background, and therefore *detection* is enhanced. On the other hand, *classification* or at least those features used by classification which rely on the fidelity of the profile, will suffer to an extent that depends on the energies of the misplaced returns.

The likely magnitudes of possible profile extensions and highlight displacements will now be estimated for the most relevant propagation modes, i.e., surface duct in deep water and in shallow water.

9.34   Deep Water: Source and Target in Duct

The skip distance of the limiting ray (distance between surface reflections) is given by

$$\left[8\left(\frac{c}{g}\right) h_{\mathrm{d}} \right]^{1/2}$$

where

h_{d} = duct depth

c = speed of sound

g = velocity gradient

In an isothermal duct then g (pressure change only) = 17 m/s per1000 m, so

$$\text{Skip distance} = \left[8 \times \left(\frac{1500}{0.017}\right) \times 100 \right]^{1/2} = 8400 \text{ m}$$

From simple geometry, the extra path length over the direct path for the limiting ray for each surface reflection is about 2.5 m (Figure 9.23). The limiting ray gives the maximum difference as 0.3 m per km. So if the range of the target submarine is 20 km, for example, its extent could be increased by up to $20 \times 0.3 = 6$ m (and highlights misplaced by up to similar amounts along the profile). Given a large submarine (at least 50 m long, say) the extension is unlikely to seriously affect detection and echo shape recognition, but classification could suffer if it relies on profile details.

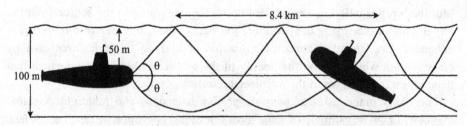

Figure 9.23   Deep water: in duct example

9.35 Deep Water: Source and Target below Duct

The bottom is remote and therefore there will only be one reflection to consider, a surface reflection (Figure 9.24). By simply approximating to a triangle, the length of the indirect path exceeds the direct path by about $2\sqrt{(R/2)^2 + h^2} - R$ (Table 9.9). Therefore, for this propagation mode, the target extension and profile distortion is greatest for deep, close targets. An extra path length of 64 m is comparable with the length of a submarine and will clearly strongly influence both echo shape recognition and classification based on profile details.

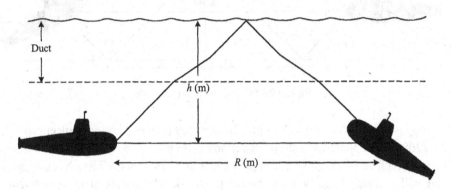

Figure 9.24 Deep water: below duct example

Table 9.9 Path difference between direct and indirect

Depth (m)	Direct path R (m)	Indirect path (m)	Difference (m)
200	10 000	10 008	8
200	5 000	5 016	16
400	10 000	10 032	32
400	5 000	5 064	64

9.36   Shallow Water: Source and Target Mid-water

Figure 9.25 simplifies the geometry. The conclusions would not differ greatly if the source were closer to the surface (e.g., in a surface ship dome). Two bottom reflections are shown but in reality there will be many more; the longest indirect path will have the most and result from the largest angle of depression, θ. Angle θ is also the grazing angle for the reflections.

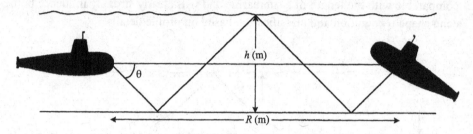

Figure 9.25   Shallow water example

Suppose $\theta = 10°$ and $h = 200$ m. The distance between bottom reflections will be 2268 m. For the two bottom reflections in Figure 9.25, $R = 3403$ m and the indirect path is 3455 m. The path difference is 52 m. This is already quite large and will increase linearly with the range, R. Bottom losses at each reflection, however, reduce the amplitudes of the multipath arrivals. (Surface reflection losses are much smaller and are neglected.)

Table 9.10 shows differences in path lengths for different angles at representative ranges of around 11 km and 23 km. Note that the large differences are accompanied by large total reflection losses and can therefore be neglected. Furthermore, bottom reflection losses increase rapidly at grazing angles greater than about 10°, so even if vertical half-beamwidths are greater than 10°, there should be no need to consider rays at angles greater than this.

Table 9.10   Path length differences at different angles

θ	Direct path, R (m)	Indirect path (m)	Difference (m)	Bottom reflections	Loss per reflection (dB)	Total loss (dB)
1°	11 458	11 460	2	1	1	1
4°	11 449	11 464	15	2	2	4
9°	11 365	11 506	141	5	3	15
1°	22 916	22 920	4	1	1	1
4°	22 898	22 928	30	3	2	6
9°	22 730	23 012	282	9	3	27

9.37 Problems

9.1 A shipborne sonar transmits a pulse at 5 kHz and receives an echo at 5.025 kHz from a target bearing G45°. If the ship is moving at 8 knots, what is the doppler, or relative velocity, of the target?

9.2 An LPFM pulse has a bandwidth of 400 Hz centred on 4000 Hz. Given own doppler nullification, what are the target dopplers for processing losses of 1 and 3 dB?

9.3 An active sonar uses a Hamming shaped CW pulse to improve the reverberation rejection, R_j, due to target doppler. How much target doppler would be needed to achieve $R_j = -40$ dB for a pulse of 1 s duration at 3000 Hz, 1000 Hz and 300 Hz?

10

Echo Sounding and Side Scan Sonars

10.1 Common Features

Both echo sounding and side scan sonars use the motion of the sonar platform to build up a representation of the sea bed and targets between the source and the sea bed. The operating frequencies range between about 20 kHz and 500 kHz – the frequency, as always, determined by the requirements of range (depth) and target size.

10.2 Echo Sounders

The first echo sounders were used simply to measure the depth of water below the sonar platform and replaced the traditional line and lead. Modern echo sounders, with narrower beams and improved displays, are also used to detect schools of fish below the sonar platform. For this application the echo sounder is often attached to the trawl, enabling the trawlerman to control the depth of his trawl very accurately; and because the distance to the sea bed is now the height of the trawl above it, the trawler can avoid damage through contact with the bottom. When attached to the trawl, the device is known as a netsonde

A typical echo sounder might operate at 100 kHz and, for a depth scale of 200 m, transmit a pulse four times per second. The standard display presentation is known as an *echogram*. Depth (strictly, distance measured from the transducer)

is on the vertical axis, and time, or distance travelled over the sea bed, is on the horizontal axis.

The echogram in Figure 10.1 displays some 30 ping returns and therefore, at a ship's speed of 10 knots (5m/s) the distance travelled to build up the echogram is about 40 m. The school of fish can be seen to extend about 5 m in plan, and about 40 m in depth. The brightness or colour of the display will indicate the strength of the echoes.

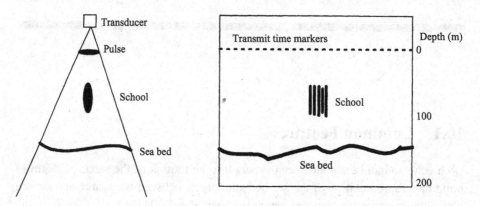

Figure 10.1 Echo sounder and display

There are no fundamental differences between the design of echo sounding and side scan sonars and the design of a submarine detection sonar. The differences are only of scale – frequency, range and target size. Because we are here concerned only with short-range or vertical transmissions, refraction of sound, which severely limits the performance of long-range antisubmarine sonars, is unimportant and the propagation losses are well described by spherical spreading plus absorption.

10.3 Echo Sounder: Design Example

The correct choice of operating frequency leads easily into a straightforward design procedure. A typical operating frequency for an echo sounder for use over continental shelf areas, at depths up to 500 m, is 40 kHz. In the deep oceans, lower frequencies will be needed to avoid excessive absorption losses which would otherwise limit the maximum depth; 20 kHz is probably about the highest practical frequency. In freshwater lakes where the absorption losses are very much less, or to detect small targets like plankton close to the surface, higher frequencies may be used, typically 100–500 kHz.

Let us begin with an outline specification (italic in Table 10.1), for an echo sounder for continental shelf areas, and then build up a more complete specification.

Table 10.1 Echo sounder: outline specification

Maximum depth	*500 m*
Frequency	*40 kHz*
Beamwidth, solid angle	*5°*
PL	59 dB at 500 m
	41 dB at 100 m
DI, transmit and receive	26 dB
Pulse length, T	133 µs
$10 \log T$	−39 dB
Ambient sea noise, N	35 dB
TS, sea bed	+10 dB
TS, individual fish	−50 dB
Source level, SL	187 dB

The propagation loss will be given by spherical spreading plus absorption (quite accurately in this instance, due to the lack of boundaries, rather than the usual working approximation).

$$PL = 20 \log r + \alpha r \times 10^{-3} \quad \text{(dB)}$$

$$r = 500 \text{ m}, \ \alpha = 9 \text{ dB/km therefore PL} = 59 \text{ dB}$$

$$r = 100 \text{ m}, \ \alpha = 9 \text{ dB/km therefore PL} = 41 \text{ dB}$$

A solid angle beamwidth of 5° will result from a square array of side L, or a slightly larger diameter circular array. We have $\theta_3 = 76/Lf$, therefore $L = 76/\theta_3 f = 0.38$ m.

The DI, in both transmit and receive, is given by

$$DI = 20 \log\left(\frac{2L}{\lambda}\right) = 20 \log\left(\frac{2 \times 0.38}{0.0375}\right) = 26 \text{ dB}$$

The pulse length would have to be very short in order to resolve individual fish. Suppose the individual fish to be separated by 0.1 m, the pulse length would need to be

$$T = 2\frac{\delta R}{c} = 2 \times \frac{0.1}{1500} = 133 \text{ μs}$$

This may be too short to meet the propagation loss requirement but let us stay with it for the present. Let the background noise, N, be the ambient noise at SS2, or 35 dB at 40 kHz. Let $5 \log d = 10$ dB and $5 \log n = 3$ dB.

Then to solve the noise-limited active sonar equation for the required source level, SL, all we require is TS. The sea bed is a very strong target at normal incidence and will be quite unmistakable. We will arbitrarily assign TS = 10 dB to the sea bed. Fish, individually or in schools, will have very much smaller target strengths, say -50 dB, but the depths will be less, say 100 m where PL = 41 dB.

The noise-limited active sonar equation is

$$\boxed{2PL = SL + TS - N + DI + 10 \log T - 5 \log d + 5 \log n}$$

Therefore

$$SL = 2PL - TS + N - DI - 10 \log T + 5 \log d - 5 \log n$$

To detect the sea bed,

$$SL = 118 - 10 + 35 - 26 + 39 + 10 - 3 = 163 \text{ dB}$$

and to detect fish down to 100 m depth,

$$SL = 82 + 50 + 35 - 26 + 39 + 10 - 3 = 187 \text{ dB}$$

SL_{max} is therefore 187 dB. Is this practical? We have

$$\boxed{SL = 10 \log P + 170.8 + DI_t}$$

Therefore

$$10 \log P = 187 - 171 - 26 = -10 \text{ dB}$$

giving $P = 100$ mW.

The acoustic power intensity is $10^{-1}/0.38^2 = 0.7 \text{W/m}^2$ and, clearly, cavitation will not be a problem. The array design is therefore practical and the remaining parameters may be inserted in Table 10.1.

The echo sounder or fish finder would in practice have a selection of source levels and pulse lengths under operator control. It would clearly be practical to have a source level much greater than 187 dB – a value of 207 dB would still only need 10 W of acoustic power.

The broadband, FM, techniques essential for the detection of many larger targets at longer ranges, and offering excellent performance against a background of reverberation, are not essential for these systems. The necessary range resolution has been shown to be obtainable using a very short CW pulse, and fish detection is normally against a noise background only; the major source of reverberation is the sea bed, which is beyond the range of most fish species of interest. An exception is the volume reverberation due to dense concentrations of plankton, which can limit the detection of individual fish.

10.4   Side Scan Sonar

A side scan sonar builds up a two-dimensional picture of the sea bed – together with any targets on the sea bed – using a combination of an asymmetrical transducer and the motion of the sonar platform through the water. The transducer may be mounted on the keel of a ship or be located in a towed body. A towed body will have the advantage of depth capability (i.e., it may usefully be placed closer to the sea bed). The installation, particularly if it is located on the keel of a ship, will often include an echo sounder. Figure 10.2 shows the vertical beam coverage provided by two transducers:

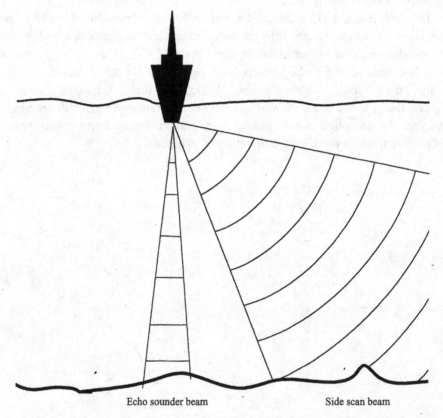

Echo sounder beam   Side scan beam

Figure 10.2   'Vertical' beams: combined echo sounder and side scan sonar

- An echo sounder transducer pointing directly downwards with a solid beam-width of 5°.

- A side scan transducer having a 'vertical' beamwidth of 45° and a 'horizontal' (fore and aft) beamwidth of 2°.

As the transmissions from the side scan transducer glance across the sea bed, each ping builds up a single line on the display (Figure 10.3). As the ship moves, therefore, successive pings build up a two-dimensional map of the sea bed. If the bottom is smooth, the display will simply show a characterless, noise-like picture. If, however, the sea bed has features, such as peaks and valleys, the picture will be quite different; the peaks will backscatter strongly and the valleys, shielded by the peaks, will display as shadows. Objects on the sea bed – wrecks, mines and bottomed submarines – are frequently detected by their shadows. This mechanism, the so-called *shadowgraph effect*, is very important for the detection and classification of bottom mines. Because of their very low target strengths, echoes from mines are very difficult to detect against a background of bottom reverberation; the shadowgraph can, however, indicate the presence of a mine even when its echo is hidden by the reverberation.

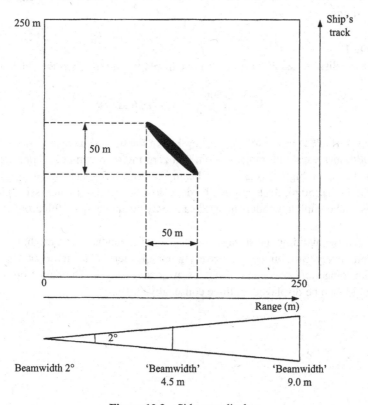

Figure 10.3 Side scan display

   A true geographic representation of the sea bed or a target requires that the scales of both display axes shall be the same. This is achieved by choosing a platform speed, V, which will match the track dimension to the maximum range dimension of the selected range scale:

$$V = \frac{R_s}{n \times \text{PRI}} = \frac{c}{2n}$$

where

   R_s = selected range scale

   n = number of pixcells [pixels?] or lines in the track dimension

$\text{PRI} = 2R_s/c$

But because the horizontal beamwidth is finite, albeit small, the beam patterns from successive pings could overlap significantly (depending on PRI and ship's speed) and therefore the data on successive lines may not be independent, except at close range.

Example 10.1
Suppose the display has 250 lines, each one displaying a ping's worth of data, then

$$V = \frac{1500}{2 \times 250} = 3 \text{ m/s (6 knots)}$$

The pings or lines are therefore spaced by 1 m in the track dimension. Since the linear beamwidth increases with range – from 0 at zero range to 9 m at 250 m range for the $2°$ wide beam in Figure 10.3 – the beam patterns from successive pings overlap significantly; at 250 m, for example, the data from 8, i.e., $9 - 1$, successive pings fully overlaps, and completely independent data is only available every 9th ping.

A compromise, avoiding the storage and display of redundant data, might be to use the data from every 3rd ping, thus ensuring independent data at about one-third of maximum range and reasonable overlaps at greater ranges. The data from every 3rd ping would then be displayed on three consecutive lines.

10.5   Side Scan Sonar: Design Example

Let us again begin with an outline specification (italic in Table 10.2) for a side scan sonar for continental shelf areas, and then build up a more complete specification. The transducer is mounted in a towed body and therefore, because it can be quite close to the sea bed, the maximum range requirement is 200 m. Let the frequency be 100 kHz and the combined (transmit and receive) beamwidth 1° fore and aft (horizontal) and 45° in the vertical plane normal to the track of the towed body.

Table 10.2   Side scan sonar: design example

Maximum range	*200 m*
Frequency	*100 kHz*
Combined beamwidth	
fore and aft	*1°*
normal to track	*45°*
PL	*53 dB at 200 m*
DI, transmit and receive	21 dB
Pulse length, T	133 µs
$10 \log T$	−39 dB
Ambient sea noise, N	30 dB
TS, sea bed	−40 dB
TS, individual fish	−50 dB
Source level, SL	204 dB

The propagation loss will be approximated by spherical spreading plus absorption (not quite so accurately in this instance, because the sea bed presents one boundary after the wavefront reaches it).

$$PL = 20 \log r + \alpha r \times 10^{-3} \quad \text{(dB)}$$

$$r = 200 \text{ m}, \ \alpha = 35 \text{ dB/km therefore PL} = 53 \text{ dB}$$

Transmit and receive on separate arrays of identical size and tilted downwards by 45°. To achieve the required combined beamwidths, each array must have double the combined beamwidths and the dimensions are given by $L = 76/f\theta_3$. Therefore, the width is $76/(100 \times 2) = 0.38$ m and the height is $76/(100 \times 90)$

= 0.0084 m. Hence each array will have two rows of 50 elements, all spaced by $\lambda/2$.

DI, in both transmit and receive, is given by

$$DI = 10\log(4Lh/\lambda^2)$$

$$= 10\log[(4 \times 0.38 \times 0.0084)/0.0152] = 18 \text{ dB}$$

Baffling will add 3 dB to this, making DI = 21 dB. Once again, the pulse length would have to be very short in order to resolve individual fish. Suppose the individual fish to be separated by at least 0.1 m; the pulse length would need to be

$$T = \frac{2 \times \delta R}{c} = \frac{2 \times 0.1}{1500} = 133 \text{ μs}$$

This is probably shorter than necessary to map the sea bed or to display submarines and wrecks but it may well meet the propagation loss requirement, so let us stay with it for the present.

Let the background noise, N, be the ambient noise at SS4, equal to 30 dB at 100 kHz, and let $5\log d = 10$ dB and $5\log n = 3$ dB. Then to solve the noise-limited active sonar equation for the required source level, SL, all we require is TS. Backscattering from the sea bed at high grazing angles will be strong and its target strength will be given by

$$TS_R = S_b + 10\log A \quad \text{(dB)}$$

where

$$A = \frac{cT}{2} R\theta_h \quad (\theta_h \text{ in radians})$$

$$= (750 \times 0.000\,133)(200 \times 0.017)$$

$$= 0.34 \text{ m}^2$$

Take the backscattering strength, S_b, to be −35 dB at the maximum range (smallest grazing angle), then

$$TS_R = -35 + 10\log 0.34 = -40 \text{ dB}$$

At shorter ranges, the reverberating area will be smaller but the backscattering strength will be greater (higher grazing angle), therefore the TS_R may not change much. This is very much smaller than the TS of 10 dB used in the echo sounder design, but remember that was for a normal incidence reflection rather than backscattering. Again fish, individually or in schools, will have small target strengths, say −50 dB.

The noise-limited active sonar equation is

$$2PL = SL + TS - N + DI + 10 \log T - 5 \log d + 5 \log n$$

Therefore

$$SL = 2PL - TS + N - DI - 10 \log T + 5 \log d - 5 \log n$$

To detect the sea bed,

$$SL = 106 + 40 + 30 - 21 + 39 + 10 - 3 = 201 \text{ dB}$$

and to detect fish out to 200 m range,

$$SL = 106 + 50 + 30 - 21 + 39 + 10 - 3 = 211 \text{ dB}$$

SL_{max} is therefore 211 dB. Is this practical? We have

$$SL = 10 \log P + 170.8 + DI_t$$

Therefore

$$10 \log P = 211 - 171 - 21 = 19 \text{ dB}$$

giving $P = 79$ W.

The acoustic power intensity is $79/(0.38 \times 0.084) = 2.5 \text{ kW/m}^2$ and again cavitation will not be a problem. The array design is therefore practical and the remaining parameters may be inserted in Table 10.2.

10.6   Problem

10.1 An echo sounder is to be used to survey the deep ocean floor. The depth requirement is 12 km to cope with the deepest ocean trenches. Given similar parameters to Section 10.3 (but take SL = 210 dB, $T = 10$ ms and TS of the deepest ocean floor as 40 dB), what is the highest possible operating frequency? What range is achievable at 30 kHz?

11

Mine Hunting Sonars

11.1 Overview

Side scan sonars, operating at similar frequencies to the example in Chapter 10, would appear to have the necessary performance to be used for mine hunting. The short pulse lengths and narrow beams would provide sufficient discrimination in the range and bearing dimensions to detect and classify mines. Operationally, however, it is clearly preferable to detect a mine *ahead* of the ship – before it becomes a danger – and the time taken to search an area using side scanning techniques is a major liability.

A mine hunting sonar, then, is usually very similar to a hull-mounted anti-submarine sonar. The array will be housed in a keel or bow dome and, in surveillance, will have a wide arc of cover centred on the ahead bearing. Classification of mine-like contacts is difficult, particularly bottom or close tethered mines, where the background to detection and classification is frequently high discrete reverberation (clutter) from the sea bed. Target strengths are low and therefore, to discriminate against this reverberation, narrow beamwidths and short (actual or resolved) pulse lengths are required. The short pulse length may be either a CW pulse, typically less than 1 ms, or a broadband FM pulse, in which case the actual pulse length can be much longer. This will improve noise-limited performance, but now the bandwidth can also be quite large; at least 10 kHz is desirable, which gives a resolved pulse length of 0.1 ms.

Modern mines are very difficult acoustic targets; they are shaped and clad to ensure very low target strengths. In extremis, detection and classification may only be possible by closely approaching the mine using a remotely operated vehicle (ROV) housing a very similar array.

11.2   Two Broad Classes of Mine

- *In-volume or near-surface*: this class of mine must be detected against a background of noise and sea surface reverberation. In shallow water the background can also be bottom reverberation but it may be possible to avoid this by the use of vertical directivity.

- *Bottom or close tethered*: this class of mine must be detected against a background of noise and *unavoidable* bottom reverberation. It is by far the most difficult of the two and definite classification may sometimes only be possible using an ROV.

11.3   Backgrounds to Mine Detection

Figure 11.1 shows how the background to detection varies with depth and range. In zone A the background is noise only. In zone B the background is noise and sea surface reverberation. In-volume mines are detected and classified in this zone relatively easily. Bottom mines will not be detected because the wavefront has yet to reach the sea bed.

In zone C the background is noise, sea surface and bottom reverberation. Both classes of mine may be detected and classified in this zone. Except on the rare occasions when the surface reverberation is greater than the bottom reverberation, the detection and classification of *in-volume mines* can be improved by steering the beam up, as indicated by the dashed beam outline in the figure – at the range of mine 1, the wavefront of the steered beam has not reached the sea bed and therefore there will be no bottom reverberation (except through the beam sidelobes at much lower levels).

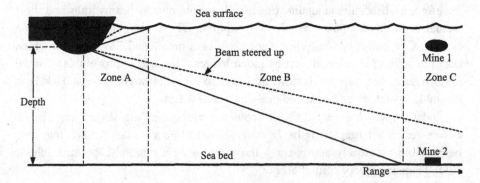

Figure 11.1   Backgrounds to the detection of mines

 A mine hunting sonar, therefore, will benefit from having more than one fan of beams available. A versatile design would include several fans, all or some of which should be displayed simultaneously. Alternatively, split-beam processing in the vertical plane may be used to achieve the same result.

11.4   Range and Bearing Resolution

A unified approach to detection and classification, as we saw in Chapter 9, demands sufficient resolution in both range and bearing dimensions to resolve the structure of both target and non-targets (false alarms) and, for mines, a bandwidth of 75 kHz was suggested, giving a range resolution of 0.01 m.

 A comparable resolution in the bearing dimension will be difficult – certainly at 'long' range, which in this context might be 100 m or more. At 100 m a 1° wide beam has a linear width of 1.7 m, acceptable for detection perhaps but too large to resolve structure for classification. The use of a broadband FM pulse and phase binning will improve the resolution; it is perhaps unrealistic to expect any better than 0.25° from the phase binning, which would give about 0.5 m resolution at 100 m and 0.05 m at 10 m. Hence the suggestion that an ROV, which can get close, might be needed for the final classification of the most difficult targets.

11.5   Design of a Mine Hunting Sonar

The sonar must be capable of detecting and classifying both classes of mine at a safe distance, at least 50 m? To ensure this, the complete system would ideally use both hull-mounted and ROV arrays. The sonar may require a choice of pulse length (certainly if only CW pulses are specified) and a choice of beamwidths, comparatively wide for surveillance but much narrower for classification. The operating frequency follows from the bandwidth requirement. We can now proceed to design the sonar.

11.6   The Threats

- *In-volume mine*: spheroid, 1 m × 0.3 m, TS = −30 dB

- *Bottom or close tethered mine*: irregular shape, minimum dimension 0.1 m, maximum dimension 1 m, TS = −30 dB

11.7   Design Example

Let us begin with an outline specification (italic in Table 11.1) for a mine hunting sonar to combat the above threats, and then build up a more complete specification. The arrays are hull mounted and the maximum slant range requirement is 300 m. The operating frequency is 200 kHz.

Table 11.1   Mine hunting sonar: outline specification

Maximum range		*300 m*
Frequency		*200 kHz*
Transmit beamwidths		
vertical		*5°*
horizontal		*90°*
Receive beamwidths		
vertical		*1°*
horizontal		*1°*
PL		*65 dB at 300 m*
Directivity index, DI		
transmit		16 dB
receive		43 dB
Source level, SL		192 dB
Pulse length, T	FM	10 ms
	CW	125 μs
Bandwidth, B	FM	80 kHz
	CW	8 kHz
$10 \log T$	FM	−20 dB
	CW	−29 dB
Ambient sea noise, N		30 dB
Reverb backscattering, S_b		−20 dB
TS		−20 dB
$5 \log d$		10 dB
$5 \log n$		3 dB

To achieve narrow beamwidths and wide cover:

- *Transmit*: in azimuth over an arc of 90°, vertical beamwidth 5°

- *Receive*: horizontal and vertical beamwidths, minimum 1°

The propagation loss will be approximated by spherical spreading plus absorption (not quite so accurately in this instance, because the sea bed presents one boundary after the wavefront reaches it).

$$PL = 20 \log r + ar \times 10^{-3} \quad \text{(dB)}$$

$$r = 300 \text{ m}, \ \alpha = 50 \text{ dB/km therefore } PL = 65 \text{ dB}$$

Transmit array

To cover all depths and ranges, the transmit array must be capable of being steered in the vertical plane. It must therefore have a number of separate elements. The required number of elements will be given by $\theta = 100/n$. The transmit array is therefore a single vertical stave of 20 elements at spacing $\lambda/2$. To reduce the maximum steer needed, both transmit and receive arrays will be mounted at an angle of depression of perhaps 30°. Wavelength λ is only 7.5 mm and therefore the radiating surface is about $75 \times 4 = 300 \times 10^{-6} \text{ m}^2$. To avoid cavitation the maximum power (from Figure 1.2) is 2 kW/m².

Because of the high frequency and short pulses, this may safely be increased to, say, 10 kW/m² or $P = 3$ W. The array will be baffled and therefore $DI_t = 3 + 10 \log 20 = 16 \text{ dB}$ and

$$SL = 10 \log P + 171 + 16 = 192 \text{ dB}$$

Receive array

Again, to cover all depths, azimuths and ranges, the receive array must be capable of being steered in both the vertical and horizontal planes. The beamformer will produce fans of beams to meet the requirements of surveillance (the initial sweep) followed by a more detailed look with narrower beams for classification. The array must be large enough to form the narrowest beams; wider beams will be formed by subsets of elements or by combining beam outputs. The required number of elements will be given by $\theta = 100/n$, and since the narrowest beams are 1° in both dimensions, $n = 100$. The receive array is therefore a square of side 0.75 m. Again, the array will be baffled and therefore

$$DI_r = 3 + 10 \log(100 \times 100) = 43 \text{ dB}$$

Pulses

The major problem for a mine hunting sonar is *classification*, particularly for mines on the sea bed, which have to be detected and classified against a background of bottom reverberation. Therefore the pulses must have sufficient bandwidth to be able to discriminate against discrete reverberation (clutter). We will propose two pulse types and compare their performances:

- *FM pulse*: bandwidth 80 kHz, duration 10 ms. The pulse needs to be quite short to avoid excessive 'dead range'; a 10 ms pulse will give a dead range of 7.5 m. The target will not have any doppler, hence a linear FM pulse will suffice. (Note that the 'dead range' is not absolute; as soon as the system goes into receive mode, processing of the pulse begins but the full gain is not achieved until a complete pulse is processed, i.e., 10 ms after the start of the receive period.)

- *CW pulse*: a comparable bandwidth would imply a very short pulse $(1/80\,000 = 12.5\ \mu s)$. Such a short pulse would have negligible noise-limited performance and a compromise is necessary. Let the CW pulse have a duration of 125 μs. The bandwidth is then 8 kHz.

11.8 Performance

The noise-limited active sonar equation is

$$2\text{PL} = \text{SL} + \text{TS} - N + \text{DI} + 10\log T - 5\log d + 5\log n$$

- *FM pulse*: $2\text{PL} = 192 - 30 - 30 + 43 - 20 - 10 + 3 = 148$ dB therefore PL $=$ 74 dB. Using this pulse, the requirement for PL $= 65$ dB is easily met.

- *CW pulse*: $2\text{PL} = 192 - 30 - 30 + 43 - 39 - 10 + 3 = 129$ dB therefore PL $=$ 65 dB. Using this pulse, the requirement for PL $= 65$ dB is now only just met.

The reverberation-limited active sonar equation is

$$10\log R = 10\log(B/\theta_h) - S_b - 11 + \text{TS} - 5\log d + 5\log n$$

When R is in metres, the constant is 11 dB.

- *FM pulse*: $10\log R = 49 + 30 - 11 - 30 - 10 + 3 = 31$ dB therefore $R = 1300$ m. The reverberation-limited range requirement is easily met with this pulse.

- *CW pulse*: $10\log R = 39 + 30 - 11 - 30 - 10 + 3 = 21$ dB therefore $R = 130$ m. The reverberation-limited range requirement is not met with this pulse.

Note that a shorter CW pulse, about 50 μs, would meet the reverberation-limited range requirement, but not the noise-limited range requirement. More importantly, however, even a 50 μs duration CW pulse has poorer range resolution than the FM pulse. We therefore conclude:

- The best solution for the composite detection and classification task is to use the FM pulse.

If this conclusion appears to conflict with some in-service mine hunting sonars, it is because of technological limitations when they were designed.

11.9   Classification

Classification demands the best possible discrimination in both range and azimuth. In the range dimension, the resolved pulse length for the FM pulse is 12.5 µs, giving *a range resolution* of about 9 mm. This is, of course, very small compared to the dimensions of a mine and there will be many returns in the range dimension to help with classification.

In azimuth, the linear resolution is proportional to range – at 100 m the resolution is 1.7 m for the 1° wide receive beams. Only at very close ranges is the resolution in azimuth comparable with the range resolution. Phase binning can improve this, but realistically not more than a factor of 4 (0.25°). So the best we can expect at 100 m is an *azimuth resolution* of about 0.5 m. This resolution, although quite large compared with the dimensions of most mines, means that, dependent upon aspect, several returns from different bearings are still possible from larger mines every ping.

An important aid to the classification of mines is the *acoustic shadow* projected onto the sea bed by bottom objects. The effect is illustrated in Figure 11.2. At the beginning of the receive period, the background is just noise. When the lower edge of the vertical beam intercepts the sea bed, reverberation adds to the noise as the pulse sweeps along the sea bed. Once the pulse has reached the mine then echoes – which may not exceed the background – will be added to the background. The

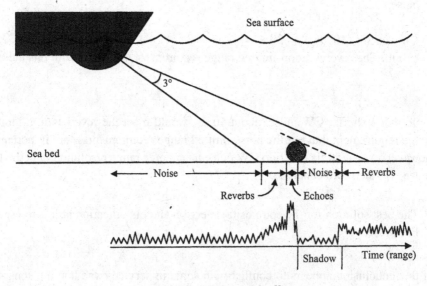

Figure 11.2   Shadowgraph effect

presence of the mine interrupts the bottom reverberation, resulting in a 'shadow' where the background is noise only. At the end of the shadow range, bottom reverberation returns until possibly interrupted again by another mine or other sources of discrete reverberation. Note that the shadow can still be produced by a mine or other source which is not strong enough to be detected from its echoes. Shadows can therefore betray the presence of mines possessing insufficient TS for normal detection from echoes. A shadow will only occur if the target extends (in azimuth) across all or most of the beam or phase bin width.

11.10   Mine Avoidance

Mine hunters are specialized vessels dedicated to the location and clearance of mines. Other vessels will clearly benefit from the provision of a sonar that can detect at least in-volume or near-surface mines so that avoiding action may be taken. Bottom mines are far more difficult to detect and classify, and a *mine avoidance* sonar will almost certainly not detect the worst of them; if bottom mines are suspected, a *mine hunting* sonar will normally have to be used.

A mine avoidance sonar will differ from a mine hunting sonar primarily in its range and azimuth cover requirements: the fitted vessel must detect mines sufficiently early to take avoiding action, and the maximum range of a mine hunting sonar is more akin to the minimum range of a mine avoidance sonar.

It is possible to adapt a sonar whose primary function is to detect submarines for the role of mine avoidance. It would employ an FM pulse using all the available transducer bandwidth and having a shorter than usual duration to avoid too long a dead range. A useful performance against in-volume and near-surface mines is possible, but because of the limited bandwidth available, bottom and close tethered mines will be very difficult to detect, let alone classify, at avoidable ranges. Furthermore, the TS of a mine falls with reducing frequency (the mine is smaller in terms of the wavelength), although this is partially compensated by the backscattering strength of bottom reverberation also falling with reduced frequency. Steering the beams in the vertical dimension will again be useful in an attempt to ensure that the background to detection is noise only. The resulting design will not be optimum and the preferred approach is to use a dedicated sonar designed for this task.

The following two sonars – one an adaptation and the other a dedicated design – will indicate the performances achievable from the two approaches.

11.11   Mine Avoidance Sonars

11.11.1   Adaptation of an antisubmarine sonar

The hypothetical sonar used for comparisons between full- and half-beam processing will be used again here. Because we need the smallest possible beamwidths, phase binning is used and a CW pulse will not be considered. The bandwidth available from the omnidirectional projector is assumed to be half an octave, or 5 kHz. Table 11.2 shows the parameters, modified where appropriate (modified parameters are in italics).

Table 11.2   Adapting an antisubmarine sonar

Parameter	Value	Units
Frequency	10	kHz
Pulse length, T	*100*	*ms*
Source level, SL	210	dB
FM bandwidth	*5000*	*Hz*
Beamwidth, full aperture	8	degrees
Beamwidth, half-aperture	16	degrees
Phase bin width	*1*	*degrees*
RI for FM with half-beams	*37*	*dB*
DI (receive)	12	dB
Target strength, TS	*−30*	*dB*
$5 \log d$	10	dB
$5 \log n$	3	dB
Background noise level	50	dB
S_b	*−30*	*dB*

Modified parameters are in italics

Detection performance

The noise-limited performance is given by

$$\boxed{2\text{PL} = \text{SL} + \text{TS} - N + \text{DI} + 10 \log T - 5 \log d + 5 \log n}$$

and using the parameter values of Table 11.2, we have

$$2\text{PL} = 210 - 30 - 50 + 12 - 10 - 10 + 3 = 125 \text{ dB}$$

If we assume spherical spreading and absorption, the range (from Figure 3.3) is 1000 m.

The reverberation-limited performance is given by

$$10 \log R = \text{RI} - S_b - 11 + TS - 5 \log d + 5 \log n$$

and using the parameter values of Table 11.2, we have

$$10 \log R = 37 + 30 - 11 - 30 - 10 + 3 = 19 \text{ dB}$$

The reverberation-limited range is only 80 m. So a mine with $TS = -30$ dB *against a reverberation background of* $S_b = -30$ dB will not be detected at a realistic avoidance range. This is quite a high reverberation background but it does show the advantage of being able to steer the beams upwards to avoid bottom reverberation. In-volume and near-surface mines can then be detected against a background of sea surface reverberation that is unlikely to have a backscattering strength of more than -40 dB. The reverberation-limited range would then be 800 m, comparable with the noise-limited range. The most difficult bottom mines may well have target strengths even less than -30 dB and therefore, even given a benign reverberation background ($S_b = -40$ dB, say), medium frequency sonars are unlikely to be able to detect all mines on or close to the sea bed. Furthermore, the poorer resolution, implicit in the smaller bandwidth, will make the classification of all mines more difficult.

11.11.2   Dedicated design

The range necessary for reliable mine avoidance is unlikely to be achieved at the very high frequencies used for mine hunting. The frequency chosen must, however, be high enough to provide sufficient bandwidth for a good reverberation-limited performance. We have already seen that an FM pulse gives superior performance in all environments. (A CW pulse can only be better given target doppler, which is not normally the case with mines.)

Let us begin with an outline specification (italic in Table 11.3), for a mine avoidance sonar to combat the threat posed by a mine of $TS = -30$ dB, and then build up a more complete specification. The arrays are hull mounted and the maximum range requirement is 1000 m. The operating frequency is 80 kHz and the bandwidth 40 kHz (half an octave).

Table 11.3   Mine avoidance sonar: outline specification

Maximum range	*1000 m*
Frequency	*80 kHz*
Bandwidth	*40 kHz*
Transmit beamwidths	
vertical	*20°*
horizontal	*90°*
Receive beamwidths	
vertical	*2°*
horizontal	*2°*
PL	*83 dB at 1000 m*
RI (= $10 \log(40\,000/2)$	43 dB
Directivity index, DI	
transmit	10 dB
receive	45 dB
Source leve, SL	193 dB
Pulse length, T	100 ms
$10 \log T$	−10 dB
Ambient sea noise, N	35 dB
TS	−30 dB
$5 \log d$	10 dB
$5 \log n$	3 dB

To achieve narrow beamwidths and wide cover:

- *Transmit*: in azimuth over a forward arc of 90°, vertical beamwidth 20°
- *Receive*: horizontal and vertical beamwidths, minimum 2°

The propagation loss is calculated using spherical spreading plus absorption (not very accurately because the sea surface and the sea bed will present boundaries for much of the ping).

$$PL = 20 \log r + ar \times 10^{-3} \quad \text{(dB)}$$

$r = 1000$ m, $\alpha = 23$ dB/km therefore PL = 83 dB

Transmit array

The height of the transmit array is given by $\theta = 20° = 100/n$. The transmit array is therefore a single vertical stave of 5 elements at spacing $\lambda/2$ and mounted at an angle of depression of 10°. Wavelength λ is only 18.75 mm and therefore the radiating surface is about $300 \times 5 = 1500 \times 10^{-6}$ m^2 (i.e., 5 circular elements of diameter 18.75 mm). To avoid cavitation the maximum power (from Figure 1.2) is 2 kW/m^2. Because of the high frequency and the short pulses, this may safely be increased to, say, 10 kW/m^2 or $P = 15$ W. The array will be baffled and therefore $DI_t = 3 + 10 \log 5 = 10$ dB and

$$SL = 10 \log P + 171 + 10 = 193 \text{ dB}$$

Receive array

To cover all depths and azimuths, the receive array is an incomplete cylinder made up of a number of staves spaced at $\lambda/2$ (Figure 11.3). Each stave will have $100/2 = 50$ elements. The diameter of the array is given by $\theta_h = 88/df$. Therefore diameter $d = 88/(2 \times 80) = 550$ mm. Each full beam, or codirectional pair of half-beams, is produced using 120° of the array. To provide 90° of azimuth cover, the staves must occupy 240° around the periphery of the cylinder. There will therefore be a total of $(\pi \times 550 \times 2)/(3 \times 9.375) = 120$ staves. The ahead beam, for example, will use the 60 staves in segments B and C. Adjacent beams are formed by stepping around the array by one stave (the staves have an angular spacing of 2°). Several fans of beams in the vertical dimension will be formed to cover the required depths and range. The near-horizontal fans will avoid bottom reverberation until quite long ranges, well beyond avoidance range.

The directivity index of the array is given by the formula for a cylindrical array used previously:

$$DI_r = 10 \log(5hdf^2) = 10 \log(5 \times 0.9375 \times 0.55 \times 80^2) = 42 \text{ dB}$$

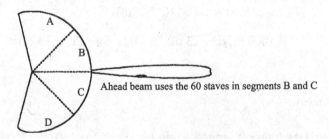

Figure 11.3   Receive array: plan view

FM pulse

The FM pulse has bandwidth 40 kHz and duration 100 ms. The pulse needs to be quite short to avoid excessive 'dead range'; a 100 ms pulse will give a dead range of 75 m. The target will not have any doppler, hence a linear FM pulse will do.

Detection performance

The noise-limited performance is given by

$$2\text{PL} = \text{SL} + \text{TS} - N + \text{DI} + 10 \log T - 5 \log d + 5 \log n$$

and using the parameter values of Table 11.3, we have

$$2\text{PL} = 193 - 30 - 35 + 42 - 10 - 10 + 3 = 153 \text{ dB}$$

giving PL = 77 dB and the range is about 800 m. To achieve 1000 m, PL must be 83 dB, i.e., we need another 6 dB. Several possibilities are open to us:

- The noise level is SS4; if we had taken SS2, the noise level would have been 30 dB.

- The pulse length could be increased (but watch the dead range); increasing T to 300 ms would increase PL by 2.5 dB.

- Increasing SL by 10 dB should be possible; the receive DI is already high (large aperture array). A selection of these changes should provide the extra decibels. Note that because of the 2PL in the equation, an extra 12 dB is needed to gain 6 dB in PL.

The reverberation-limited performance is given by

$$10 \log R = \text{RI} - S_b - 11 + \text{TS} - 5 \log d + 5 \log n$$

and using the parameter values of Table 11.3, we have

$$10 \log R = 43 + 30 - 11 - 30 - 10 + 3 = 25 \text{ dB}$$

The reverberation-limited range is therefore 316 m. So *against a reverberation background of* $S_b = -30$ dB, a mine with TS $= -30$ dB is detected at a range where avoidance is still possible, but less than the noise-limited range of 800–1000 m. This is quite a high reverberation background but again it shows the advantage of being able to steer the beams upwards to avoid bottom reverberation. In-volume and near-surface mines can then be detected against a background of sea surface reverberation, which is unlikely to have a backscattering strength of more than -40 dB. The reverberation-limited range would then be 3160 m, much greater than the noise-limited range. The most difficult bottom mines may well have target strengths even less than -30 dB, so even a dedicated mine avoidance sonar will find it difficult to detect all mines on or close to the sea bed.

Classification

Classification demands the best possible discrimination in both range and azimuth. In the range dimension, the resolved pulse length for the FM pulse is 25 µs, giving a *range resolution* of about 18 mm. This is very small compared to the dimensions of a mine, and there will be many returns in the range dimension to help with classification.

In azimuth the linear resolution is proportional to range. Classification can be at a closer range than initial detection; if we assume 300 m, the resolution is 10 m for the 2° wide receive beams. The resolution in azimuth is never remotely comparable with the range resolution. Phase binning can improve this, but realistically not more than a factor of 8 (0.25°). So the best we can expect at 300 m is an *azimuth resolution* of about 1.3 m. This resolution is comparable with the dimensions of most mines, therefore the majority of returns from a mine will be at the same bearing and a complete picture of a mine will not be available. However, even this resolution is useful for classification because it will discriminate against large discrete patches of bottom reverberation, which can extend over much greater dimensions.

11.12 Problems

11.1 How would the mine hunting sonar of Section 11.7 perform if TS were -25 dB?

11.2 The mine avoidance sonar design of Section 11.11.2 took the PL at the centre frequency of the pulse (80 kHz). At the highest frequency of the pulse, what SL is necessary to achieve a range of 800 m? And assuming SL does not change with frequency over the bandwidth of the pulse, what equalization is needed to maintain a flat response at the receiver?

12

Intercept and Communications Sonars

12.1   Intercept Sonars

An intercept sonar detects active transmissions from other sonar platforms – a submarine intercept sonar, for example, will have to be capable of detecting and classifying active transmissions from surface ships, other submarines and helicopter dipping sonars, and the active transmissions from torpedoes in their homing stages.

An intercept sonar, unlike an active sonar, will have no a priori knowledge or incomplete a priori knowledge of the signals it has to detect. Hence matched filter processing cannot – at least initially – be used and the basic intercept receiver simply employs *energy detection*.

The arrays and processor will have to accept a wide frequency spectrum – from about 1 to 20 kHz for search sonar transmissions, and from about 20 to 100 kHz for weapon homing transmissions. The arrays will not be optimum throughout this band and typically the DI might be between 3 and 10 dB.

Nevertheless, because the intercepted pulse has only undergone a *one-way* propagation loss – from source to receiver – compared with the *two-way* propagation loss of the intercepted active sonar, intercept ranges will be at least equal to, and typically several times greater than, the ranges achieved by the active sonar.

12.2   Communications Sonars

A communications sonar receives signals (voice or encoded messages) from another platform. But now, unlike an intercept sonar, the receiver will have an a priori knowledge of the signals it has to detect. Hence matched filter processing can now be used.

The bandwidth of the receiver will therefore only need to be wide enough to accept the frequency spectrum of the messages. Intelligible speech has a bandwidth of about 3 kHz (100–3000 Hz) but encoded messages may have much larger bandwidths, particularly if the communications channel must be covert – to be secure from intercept, the message must be fast and of short duration, which implies a large bandwidth.

As with intercept, the signals only undergo a *one-way* propagation loss, and long ranges are possible using only modest source levels.

12.3   Function of an Intercept Sonar

An intercept sonar examines detected pulses to determine the following:

- Carrier frequency

- Bandwidth

- Pulse type (FM, CW, PRN, etc.)

- Pulse duration

- Pulse shape

- Pulse interval

- Multipath structure

With the exception of multipath structure, all serve to help identify the transmitter and hence the likely platform.

The carrier frequency may be between, say, 1–100 kHz. A typical receiver to cover this spectrum might do its surveillance processing in *octave bands* (1–2 kHz, 2–4 kHz, etc.,), but should also have the capability to vary the exact bands given some a priori knowledge of likely threats.

Bandwidths may be between, say, 1 Hz (a long CW pulse) and 1000 Hz (a short CW pulse or broadband FM or PRN pulses). The intercept sonar should at least measure start and finish frequencies and ideally it would completely analyse the amplitude and frequency structure of the pulse.

Pulse interval, and any change in the interval, are important clues to the likely range of the intercepted sonar. Suppose pulse interval = 15 s (corresponding to a maximum range of about 10 km) the transmitter will be within this range, assuming it is in contact. If the pulse interval changes to say 8 s, this is a strong indication that the transmitting platform is closing range and is within 5 km of the intercept sonar.

Time delays between direct and bottom bounce arrivals in deep water can also indicate range, but again only very broadly given that the simple broadband intercept receiver is unlikely to be able to make a good measure of the vertical angle of arrival. The large planar or conformal arrays fitted to submarines, however, can measure vertical angles of arrival to sufficient accuracy to give a good estimate of range, either from time delays between direct and indirect arrivals of the pulse, or from the vertical angle of maximum response (compare with VDPR). Given a knowledge of the original pulse length of a transmission, it may be possible to infer the range of its source from any stretching of the pulse resulting from multipath. The range estimate may be no more than that if the pulse is stretched then the source is distant, and if the pulse is not stretched then the source is close.

12.4   Intercept Sonar Equation

An intercept sonar is a *passive sonar* optimized for the detection of active transmissions from other platforms. The basic *intercept sonar equation* is therefore

$$SE = (SL - PL) - N + DI - DT$$

This is the basic passive sonar equation: SL is no longer the radiated noise of a target but is the source level of the active transmission.

Because nothing can be assumed about the signal to be detected, the *detection* process for the simplest intercept sonar is based on the received signal-to-noise ratio without any processing gain, therefore

$$DT = 5 \log d + 10 \log B_r$$

Both signal and noise are referred to the same bandwidth, hence the term $10 \log B_r$ where B_r is the bandwidth of the receiver.

Combining these equations and putting $SE = 0$, we obtain the intercept sonar equation:

$$\boxed{PL = SL + DI - N - 5 \log d - 10 \log B_r}$$

12.5   Worked Examples

Example 12.1
An intercept receiver mounted on a submarine processing an octave band from 6 to 12 kHz detects a signal of bandwidth 400 Hz centred on 9 kHz. Given spherical spreading and absorption, what is the maximum range of the source of the signal, assuming it is from a sonar with an SL of 220 dB? (The submarine is aware that the sonar is likely to be Type XXX on a surface ship with a standard SL of 220 dB.)

Assume that DI $= 6$ dB at 9 kHz and background noise $N = 40$ dB:

$$PL = 220 + 6 - 40 - 10 - 10 \log 60 = 138 \text{ dB}$$

$$PL = 20 \log r + 0.9r \times 10^{-3} = 138 \text{ dB}$$

$$\text{Intercept range} = 50 \text{ km}$$

How does this compare with the detection range from the ship?

$$2PL = SL + TS - N + DI + 10 \log T - 5 \log d + 5 \log n$$

Take $N = 60$ dB, TS $= 10$ dB, DI $= 20$ dB, $T = 1$ s, $5 \log d = 10$ dB, $n = 5$:

$$2PL = 220 + 10 - 60 + 20 + 0 - 10 + 3 = 183 \text{ dB}$$

$$PL = 20 \log r + 0.9r \times 10^{-3} = 92 \text{ dB}$$

$$\text{Detection range} = 12 \text{ km}$$

Therefore the presence of the ship can be known to the submarine long before it becomes an actual threat.

Example 12.2
An intercept receiver mounted on a submarine processing an octave band from 30 to 60 kHz detects a signal of bandwidth 1000 Hz centred on 50 kHz with a ping interval of 0.5 s. What are the probable source and range of the signal?

Assume DI $= 3$ dB and $N = 30$ dB. The frequency and ping interval indicate that the source is probably a torpedo in its active homing phase. The SL may therefore be about 200 dB.

$$PL = 200 + 3 - 30 - 10 - 10 \log 30\ 000 = 118\ \text{dB}$$

$$PL = 20 \log r + 16r \times 10^{-3} = 118\ \text{dB}$$

Intercept range $= 3$ km

What is the probable detection range of the submarine by the torpedo?

$$2PL = SL + TS - N + DI + 10 \log T - 5 \log d + 5 \log n$$

Take $N = 40$ dB, TS $= 10$ dB, DI $= 20$ dB, $T = 20$ ms, $5 \log d = 10$ dB, $n = 5$:

$$2PL = 200 + 10 - 40 + 20 - 17 - 10 = 166\ \text{dB}$$

$$PL = 20 \log r + 16r \times 10^{-3} = 83\ \text{dB}$$

Detection range $= 1.3$ km

But because the ping interval is only 0.5 s (corresponding to a range scale of about 400 m), the submarine may well infer that the torpedo is well within this maximum range.

So the submarine intercept receiver is capable of detecting the torpedo active transmissions well beyond the torpedo's active acquisition range. Therefore the torpedo launch platform (submarine, surface ship, aircraft) should attempt to closely localize the target and delay homing transmissions until the torpedo is within a few hundred metres of the target.

12.6   Reduction in the Probability of Intercept

Submarines are becoming increasingly difficult to detect passively, even from other submarines, and the active capability of submarine sonars may perforce have to be used for detection purposes (rather than simply to localize a target immediately before an attack).

The active transmissions should therefore be designed for a *low probability of intercept* (LPI) and also, recognizing that this is very difficult and may not be achieved, for a *low probability of exploitation* (LPE), i.e., having detected the transmission it should be difficult for the intercept receiver to infer anything about the identity of the intercepted platform and its movements.

Unfortunately, designing for LPI is in direct conflict with designing for detection. A low probability of intercept may be achieved in several ways:

- Reducing SL

- Increasing frequency to limit range by absorption losses

- Increasing the bandwidth

- Using sector transmissions to confine the energy within azimuths of interest

If increased bandwidth is achieved by reducing the pulse length, as with CW, then detection performance will also suffer. If it is achieved without reducing the pulse length, detection performance will not suffer, but since the intercept receiver bandwidth is likely to be larger still, the performance of the intercept sonar will not suffer either.

Clearly, with the possible exception of sector transmissions, all these measures severely prejudice the detection performance of an active sonar. But a simple example will suffice to show that LPI may not be usefully improved even by the use of sector transmissions.

Example 12.3

An intercept receiver mounted on a submarine processing an octave band from 4 to 8 kHz detects a signal centred on 5 kHz from a sonar with a known sector SL of 230 dB. Given spherical spreading and absorption, what is the maximum intercept range of the sector transmission and the reduced intercept range for bearings outside the sector, for transmission sidelobes of −30 dB?

Assume DI = 6 dB at 5 kHz and assume N = 40 dB:

- Within the sector, SL is 230 dB and $PL = 230 + 6 - 40 - 10 - 10\log 4000$ $= 150$ dB. $PL = 20\log r + 0.3r \times 10^{-3} = 150$ dB therefore $r = 150$ km.

- Outside the sector, SL is 200 dB and $PL = 200 + 6 - 40 - 10 - 10\log 4000 =$ 120 dB therefore $r = 70$ km.

So the intercept range outside the sector (although only about half the range within the sector) is still very long and much greater than the probable maximum range of the intercepted sonar, perhaps 20 km. A useful goal would be to reduce the out-of-sector intercept range to this value, i.e., 20 km. What level of sidelobes would achieve this?

$$PL = 20\log 20\,000 + (0.3 \times 20) = 92 \text{ dB}$$

Therefore the sidelobes would need to be $150 - 92 = 58$ dB down. This sidelobe level is quite impractical and, even theoretically, it could only be achieved by severely compromising the sector SL and beamwidth.

Since all measures to reduce LPI compromise the performance of an active sonar, it is reasonable to conclude:

- Virtually any sonar transmission will be intercepted at a range far in excess of its own detection range.

The emphasis must therefore fall upon LPE. Note, however, that mine hunting sonars, which operate against close targets using high frequency transmissions, have a very low probability of intercept. This is because absorption severely reduces the range where the sound intensity becomes negligible.

12.7   Reduction in the Probability of Exploitation

Useful *LPE measures* will include those designed to hinder platform identification and platform movement and location. The features of an intercepted transmission examined by an intercept receiver (listed above) are exploited to *identify* the sonar and hence the likely platform types. The identification may, at best, be made non-specific by ensuring that transmissions from the different platforms within a force are as similar as possible. Once a sonar has been designed, transmitted and intercepted by an adversary, there is little that can be done to hinder any future identification of the sonar.

It is possible to infer range rate (relative velocity) from the doppler shifts to intercepted transmissions of known frequencies. The doppler shift is 0.35 Hz per knot per kHz given ODN. (Compare this with 0.69 Hz per knot per kHz for an active sonar.) A useful LPE measure, therefore, is to randomize the frequency of a pulse so that the intercept receiver infers an incorrect range rate. If, for example, the frequency of a transmission from a platform *closing* at a relative velocity of 10 knots is reduced from its nominal value of 5 kHz by 35 Hz, the intercept receiver could infer that the platform is *opening* at a relative velocity of 10 knots.

The use of a range scale greater than necessary will reduce the intercept opportunities and can confuse the intercept receiver about the range of the intercepted platform, and whether or not the active sonar is in contact (particularly if the range is known).

12.8   Effectiveness of Intercept Sonars

Intercept sonars can be exceedingly effective; they dissuade the use of active sonars – particularly by submarines with a strong desire to remain covert – because it is difficult if not impossible to avoid being detected and identified (usually long before the active sonar is in contact itself) by even a mediocre intercept receiver. Probably the best that an active sonar can do is to attempt to mislead the intercept receiver by making small random changes in transmit frequencies and making range scale changes which are at odds with the real tactical scenario.

12.9   Communications Sonars

Communications sonars are normally used for communicating between surface ships and submarines, or between submarines. They may simply use voice channels (underwater telephones) or, particularly when security and integrity are important, encoded messages. The techniques used for encoding and modulation will only be considered here insofar as they impinge on the design of the underwater communications channel itself, i.e., the *bandwidths* needed for voice and message channels.

Voice channels suffer significant distortion during underwater propagation, particularly at low frequencies (say 1–12 kHz) where the attenuation in the water can differ markedly across the 3000 Hz bandwidth required for the channel. Reverberation and multipath also add to the distortion. This distortion may be removed by digital encoding before transmission but, inevitably, at low frequencies the communications link will be slowed down. A low frequency channel has the incidental advantage that it can often use an existing search sonar array, transmitters and receivers.

Secure encoded message channels must be of short duration to reduce intercept opportunities. Therefore they need more bandwidth than a simple voice channel. A typical bandwidth is 10–50 kHz. The actual operating frequencies are correspondingly high and the range much reduced.

12.10   Communications Sonar Equation

A communications sonar may be considered as a special case of an intercept sonar. Both sonars receive active transmissions, but the communications sonar knows the nature of the signal to be detected whereas the intercept sonar does not. The *basic communications sonar equation* is therefore

$$SE = (SL - PL) - N + DI - DT$$

This is the basic passive sonar equation: SL is no longer the radiated noise of a target but is the *known* source level of the active transmission.

The *detection* process for the simplest communications sonar is based on the received signal-to-noise ratio, in the bandwidth of the communication channel, and therefore

$$DT = 5 \log d + 10 \log B_c$$

Both signal and noise are referred to the same bandwidth, hence the term $10 \log B_c$ where B_c is the bandwidth of the communication channel.

Combining these equations and putting $SE = 0$, we obtain the communications sonar equation:

$$\boxed{PL = SL + DI - N - 5 \log d - 10 \log B_c}$$

12.11   Examples of Communications Sonars

Example 12.4

An *underwater telephone* is required to use an existing surface ship search sonar array for communication with submarines. The array is tuned over a band from 8 to 10 kHz where its SL = 215 dB, and its response falls by 12 dB per octave away from resonance. The telephone is to use a band from 12 to 15 kHz. Assume that a dedicated system on the submarine will deliver the same performance. What is the SL needed for a range of 4 km?

The communications sonar equation is

$$\boxed{PL = SL + DI - N - 5\log d - 10\log B_c}$$

Let DI = 20 dB, $5\log d = 10$ dB, $N = 45$ dB (SS4 at 12 kHz, the vessels will be stationary or slow moving and the background will be ambient sea noise), $B_c = 3000$ Hz. Assume spherical spreading and absorption:

$$PL = 20\log 4000 + (1.6 \times 4) = 78 \text{ dB}$$

$$SL = 78 - 20 + 45 + 10 + 35 = 148 \text{ dB}$$

This very modest SL results from propagation being *one-way only*, and the transmitter power will need to be significantly reduced for this application of the array in order to minimize the probability of intercept.

The dedicated system on the submarine will also have modest requirements. It may only be necessary to provide a suitable projector and use an existing passive array (possibly an intercept array). The SL of the submarine array must be adjustable to minimize the probability of intercept, and in practice both vessels would reduce their transmitted powers to levels consistent with intelligible communication.

Example 12.5

An *encoded message channel* uses a band from 50 to 100 kHz. What SL is now required for a range of 4 km, the same range as in Example 12.4?

Let DI = 30 dB, $5\log d = 10$ dB, $N = 35$ dB (SS4 at 50 kHz), $B_c = 50$ kHz and assume spherical spreading and absorption:

• At 50 kHz

$$PL = 20\log 4000 + (16 \times 4) = 136 \text{ dB}$$

$$SL = 136 - 30 + 35 + 10 + 47 = 198 \text{ dB}$$

- At 100 kHz

$PL = 20 \log 4000 + (35 \times 4) = 212$ dB

$SL = 212 - 30 + 35 + 10 + 47 = 274$ dB

The large difference in SL between the ends of the band is range dependent, even if a source level of 274 dB were possible, it would be very difficult to equalize successfully. Therefore a channel of this bandwidth is only capable of much shorter ranges. A more realistic design would aim for a range of 1 km, say, then:

- At 50 kHz

$PL = 20 \log 1000 + (16 \times 1) = 76$ dB

$SL = 76 - 30 + 35 + 10 + 47 = 138$ dB

- At 100 kHz

$PL = 20 \log 1000 + (35 \times 1) = 95$ dB

$SL = 95 - 30 + 35 + 10 + 47 = 157$ dB

Now the difference in SL between the ends of the band is only 19 dB and easy to equalize.

Fortunately, a modest range is often acceptable: it is easy for a quiet submarine to detect and locate a relatively noisy surface vessel and then approach to, say, 500 m before establishing communication. Both vessels should reduce their source levels to the minima required for reliable communication (perhaps 10 dB greater than the usual values determined on detection criteria alone).

12.12   Problems

12.1 An intercept receiver mounted on a submarine processing an octave band from 10 to 20 kHz detects a CW pulse of duration 1 s at a frequency of 15 kHz. Given spherical spreading and absorption, what are the range limits of the source of the signal, assuming it is from a sonar whose SL can be varied between 180 and 220 dB? Assume that DI = 10 dB at 15 kHz and background noise $N = 40$ dB.

12.2 If the submarine has a TS = 15 dB and the DI of the intercepted sonar is 20 dB, what are the intercepted sonar's likely noise-limited ranges at these limits of SL?

13

Active Sonar Design

13.1 Introduction

This chapter will use the concepts of the previous chapters to design representative active sonar systems for the detection of submarines and torpedoes. The interactions between the initial requirements and the limitations imposed by platform size and the environment will be demonstrated by considering practical systems.

13.2 Submarine Detection

The requirement for the detection of submarines may be stated simply as: to detect and classify at beyond (the submarine's) weapon range, at an acceptable (low) false alarm rate. This requirement will, in practice, not be met for all submarines, in all environments, at all times. The designer will have to consider the constraints, primarily the platform size, and design for the best performance possible.

13.3 Hull-Mounted Surface Ship Sonar

Hull-mounted antisubmarine sonars are widely used in surface ships (and submarines). When fitted in ships, this class of sonar generally has the following characteristics:

- Cylindrical array; 1–3 m in diameter; 0.5 to 2 m in height; the size is constrained by the platform.

- Mounted in a keel dome or, more commonly for modern sonars, in a bow dome; the bow site offers the lowest self-noise.

- Operating frequencies between 3 and 15 kHz; the frequency is chosen (it almost chooses itself) to give an acceptable balance between performance against noise and reverberation backgrounds for the array size.

- Detection ranges from 5 km(15 kHz) to 20 km(3 kHz) but highly dependent on the environment.

13.4   Representative Hull-Mounted Design

The largest possible array for the ship has been determined to be 2 m in diameter. For the greatest possible noise-limited range, the operating frequency should be low. Too low, however, and the horizontal beamwidth will be too small for adequate performance against reverberation. The receive DI will also fall with reducing frequency, which will reduce the noise-limited range, thus creating an opposing requirement to increase the frequency. By iteration through a design, an optimum frequency will finally emerge. An experienced designer might start by specifying the beamwidths and proceed as follows:

$$\text{Horizontal beamwidth } \theta_h = 8°$$

$$\text{Vertical beamwidth } \theta_v = 12°$$

We have $\theta_h = 88/df_0$ therefore $f_0 = 5.5$ kHz, and we have $\theta_v = 76/hf_0$ therefore $h = 1.15$ m.

The receive DI is $10 \log 5hdf_0^2 = 25$ dB. For all-round cover use an omnidirectional transmission. The transmit DI will then be that of a line of length 1.15 m (the height of the array):

- Transmit DI $= 10 \log(2h/\lambda) = 9$ dB

- The source level is SL $= 10 \log P + 171 + \text{DI}_t$

- If we specify SL $= 224$ dB then the total acoustic power is $P = 25$ kW

The array elements should be spaced no more than $\lambda/2 = 136$ mm. Make the actual spacing 120 mm. A stave can therefore have $1150/120 = 10$ elements. (Note all dimensions are approximate and would be modified by practical considerations.)

At the crossover points of the beams there will be a 'scalloping loss' of 3 dB. To limit this loss to about 1 dB, the staves must be spaced less than a beamwidth, say $6°$, so there will be $360/6 = 60$ staves. There will therefore be a total of 600 elements in the array and, in transmit, each element must be capable of transmitting $25\,000/600 = 40$ W of acoustic power. Suppose the radiating surface of each element to be $\pi(0.05)^2$, then the total radiating surface is $600\pi(0.05)^2 = 4.7$ m^2.

The radiated power intensity is therefore $25/4.7 = 5.3$ kW/ m^2 and Figure 1.2 indicates that cavitation could be a problem unless the array is at a depth of at least 10 m. Practical arrays, however, exist and operate at about this intensity at lesser depths, so it is safe to continue with this design.

Based on the above analysis and the procedures and recommendations from previous chapters, we are now in a position to attempt to complete Table 13.1, a table of parameters for the sonar.

Table 13.1   Parameters of hull-mounted surface ship sonar

Parameter	Value	Units
Frequency	5.5	kHz
Pulse length, T	1	s
Source level, SL	224	dB
FM bandwidth	750	Hz
CW bandwidth	2	Hz
Horizontal beamwidth, θ_h	8	degrees
Reverberation index, RI		
FM	20	dB
CW	−6	dB
DI (receive)	25	dB
Target strength, TS	10	dB
$5 \log d$	12	dB
$5 \log n$	3	dB
Background noise	55	dB
S_b	−33	dB

Pulse length

The pulse length, T, has been chosen to be 1 s for both FM and CW pulses. Increasing T beyond this might increase the noise-limited range slightly but it does have operational implications – the sonar would be 'blind' out to $1500T/2$ metres, or even longer for the first pulse if both pulses are transmitted sequentially. The other argument for increasing T – improved doppler resolution and hence possibly improved performance against very low target doppler – is particularly important for a stationary platform (e.g., a helicopter dipping sonar). Chapter 9 discusses the CW target doppler problem in detail but note here that a 1 s shaped CW pulse will have a bandwidth of about 2 Hz, which corresponds to a doppler width of about 0.5 knots at an operating frequency of 5.5 kHz. This is already comparable with spreads in the reverberation background, so increasing the pulse length will not be useful.

FM bandwidth

The FM bandwidth is chosen to resolve the structure of targets and non-targets, as discussed in Chapter 9.

Reverberation indices

- For FM, $10 \log(750/8) = 20$ dB
- For CW, $10 \log(2/8) = -6$ dB

Detection index

The detection index is $5 \log d = 12$ dB. This value, for an active sonar, has $P_d = 0.9$ and $P_{fa} = 10^{-5}$. A decision after n pings, made by the operator or an automatic detector, results in an incoherent gain of $5 \log n$(dB). For $n = 5$ we have 3 dB.

Background noise

The background noise, 55 dB, is the spectrum level of sea state 4 (SS4) ambient noise at the operating frequency of 5.5 kHz. It is likely to be the dominant source of self-noise for a typical frigate at speeds of up to about 15 knots.

Target strength

The target strength (TS) is an 'average' value for a typical submarine. Should the same value be used for both pulses? It is usual for the TS to have been measured using a CW pulse of a duration which roughly matches the target extent, say 100 ms for a submarine, and it is the *peak value* which is quoted. For long CW pulses, the peak TS equals the integrated TS and the suggested value of 10 dB is therefore appropriate.

The large bandwidth of the FM pulse means that its effective pulse length is short, 1.3 ms, and the peak TS does not equal the integrated TS. The correct value to use for TS is now somewhat less (the peak TS as measured by an equivalent short pulse, which is perhaps about 5 dB less). However, we can use the same,

10 dB, value if we integrate the post-detection samples, but do not include a term in the sonar equation for this. (In other words, the signal processing does the integration and not the target.)

Performance

The noise-limited performance of the sonar is given by the following equation:

$$2PL = SL + TS - N + DI + 10 \log T - 5 \log d + 5 \log n$$

Substituting the values from Table 13.1, we obtain

$$2PL = 224 + 10 - 55 + 25 + 0 - 12 + 3 = 195 \text{ dB}$$

The allowable PL is 98 dB, and assuming spherical spreading plus absorption ($\alpha = 0.5$), the noise-limited detection range is about 22 km. This range is achieved for both the FM and CW pulses because they have the same duration.

The reverberation-limited performance of the sonar is given by the following equation:

$$10 \log R = 10 \log(B/\theta_h) - S_b - 41 + TS - 5 \log d + 5 \log n$$

Substituting the values from Table 13.1, we obtain

$$10 \log R = 20 + 33 - 41 + 10 - 12 + 3 = 13 \text{ dB}$$

and the reverberation-limited range of the FM pulse is 20 km.

But consider the effect of reducing the TS to 6 dB. The allowable PL is now 96 dB and the noise-limited detection range is about 20 km (reduced but not a dramatic change). The reverberation-limited range, however, is now only 8 km, and the sonar is reverberation limited. The design, therefore, is only well balanced given a 10 dB target and $S_b \leqslant -33$ dB.

The sonar may still, however, have a perfectly adequate performance. A value of $S_b = -33$ dB is only exceeded in about 20 per cent of UK shallow water areas and in deep water a value of $S_b = -40$ dB is high. Note that the reverberation-limited performance can also be enhanced by using half-beam processing, which effectively reduces the value of θ_h to be used in the equation for RI.

The small, 2 Hz, bandwidth of the CW pulse results in a very low RI, and its reverberation-limited range is negligible. The CW pulse is only useful given *sufficient target doppler* to ensure that detections are noise limited.

13.5 Longer Ranges

To detect submarines at ranges significantly higher than 20 km, it is necessary to use a much lower frequency. Simply to use a lower frequency and the same array size, however, would not be sufficient. In noise-limited conditions the reduced absorption losses would be more than offset by a lower receive DI and increased self-noise. For many ship designs, a 2 m diameter array is the largest practical and a different approach is required.

13.6 Towed Transmitters and Towed Array Receivers

The requirement for increased range (to counter increased weapon ranges) is often met using a *combination* of a towed body transmitter and a towed line array receiver (Figure 13.1). This combination will have a, perhaps limited, variable depth capability and operate at frequencies between, say, 300 and 2000 Hz. The receive array will be quite short compared with a passive towed array and it may be attached to the transmitting body.

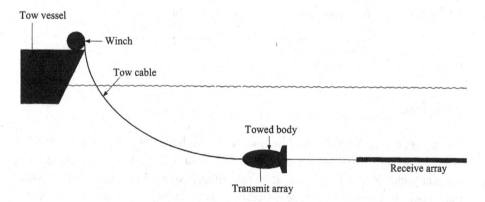

Figure 13.1   Towed active sonar system

13.7 Representative Design

Consider a low frequency active sonar operating at 1000 Hz. The transmit array comprises a vertical stack of elements of height 3 m, or 2λ, and transmitting omnidirectionally in azimuth. The source level is again 224 dB. The vertical beamwidth is $\theta_v = 50\lambda/L = 25°$. The receive line array is 32λ in length, which is 48 m at 1000 Hz. At broadside the horizontal beamwidth is $\theta_h = 50\lambda/L = 1.6°$; θ_h increases at angles away from broadside, as shown in Figure 2.2. If we define operational bearings as $\pm60°$, then the maximum value for θ_h is approximately $3°$. The receive DI is $10\log n = 18$ dB. Once again we can complete a table of parameters for the sonar (Table 13.2), based on the above and the procedures and recommendations from previous chapters.

Table 13.2 Parameters of low frequency active sonar

Parameter	Value	Units
Frequency	1	kHz
Pulse length, T	2	s
Source level, SL	224	dB
FM bandwidth	300	Hz
CW bandwidth	1	Hz
Horizontal beamwidth, θ_h	3*	degrees
Reverberation index, RI		
FM	20	dB
CW	−2	dB
DI (receive)	18	dB
Target strength, TS	10	dB
$5\log d$	12	dB
$5\log n$	3	dB
Background noise	70	dB
S_b	−33	dB

*Maximum

Pulse length

The pulse length, T, has been chosen to be 2 s for both FM and CW pulses. A 2 s shaped CW pulse will have a bandwidth of about 1 Hz, which corresponds to a doppler width of about 1.5 knots at an operating frequency of 1 kHz. This is larger than possible spreads in the reverberation background, so increasing the pulse length of the CW pulse may be useful. Note well that as frequency is reduced, there needs to be a greater target doppler (relative velocity in knots) to achieve noise-limited performance with a CW pulse.

FM bandwidth

The FM bandwidth is now limited to 300 Hz by the available bandwidth for an acceptable projector transmit efficiency. This is less than optimum to resolve the structure of both targets and non-targets, but it is acceptable.

Reverberation indices

- For FM, $10 \log(300/3) = 20$ dB
- For CW, $10 \log(1/3) = -5$ dB

Target strength

The target strength (TS) is an 'average' value for a typical submarine.

Detection index

The detection index is $5 \log d = 12$ dB. This value, for an active sonar, has $P_d = 0.9$ and $P_{fa} = 10^{-5}$. A decision after n pings, made by the operator or an automatic detector, results in an incoherent gain of $5 \log n$(dB). For $n = 5$ we have 3 dB.

Background noise

The background noise, 70 dB, is the spectrum level of sea state 4 (SS4) ambient noise at the operating frequency of 1 kHz. The self-noise of a towed array is that of the ambient sea noise at towing speeds up to perhaps 10 knots, except at ahead bearings where the radiated noise of the tow ship may be dominant (see Example 8.2).

Performance

The noise-limited performance of the sonar is given by the following equation:

$$2PL = SL + TS - N + DI + 10 \log T - 5 \log d + 5 \log n$$

Substituting the values from Table 13.2, we obtain

$$2PL = 224 + 10 - 70 + 18 + 3 - 12 + 3 = 176 \text{ dB}$$

The allowable PL is 88 dB, and assuming spherical spreading plus absorption ($\alpha = 0.06$), the noise-limited detection range is about 26 km. This range is achieved for both the FM and CW pulses because they have the same duration. This is significantly greater than the 22 km range of the hull-mounted sonar of the previous example, operating at a frequency of 5.5 Hz. Note that it has been achieved at a much lower PL (88 dB compared with 98 dB). This is due to the much smaller absorption losses at 1 kHz.

The reverberation-limited performance of the sonar is given by the following equation:

$$10 \log R = 10 \log(B/\theta_h) - S_b - 41 + TS - 5 \log d + 5 \log n$$

Substituting the values from Table 13.2, we obtain

$$10 \log R = 20 + 33 - 41 + 10 - 12 + 3 = 13 \text{ dB}$$

and the reverberation-limited range, using the FM pulse, is 20 km. It is the same as the hull-mounted sonar because the RI is unchanged. (The smaller bandwidth is compensated by the narrower horizontal beamwidth.)

The small, 1 Hz, bandwidth of the CW pulse results in a very low RI. and its reverberation-limited range is negligible. As always the CW pulse is only useful given sufficient target doppler to ensure that detections are noise limited. The remarks on reverberation-limited performance made above for the hull-mounted sonar apply equally here.

Here are some ways we could use to improve the noise-limited performance of this sonar:

- Double the length of the towed array = +3 dB

- Increase SL = +3 dB

- Transmit a train of five 2 s pulses
  and incoherently add ($5 \log 5$) = +3 dB

This is a total increase in 2PL of 9 dB, and PL becomes 93 dB. The noise-limited range is now 35 km. Note that a similar increase in 2PL for the hull-mounted sonar (which is probably only possible by resorting to sector transmissions) would have increased its range to about 28 km.

A useful insight into the performance of any active system is obtained by plotting the levels of echo, noise and reverberation against range:

$$\text{Echo level} = EL = SL + TS - 2PL$$

$$\text{Reverb level} = RL = SL + (S_b + 10 \log A) - 2PL$$

$$\text{Noise level} = NL = N - 10 \log T - DI$$

Using the parameters from Table 13.2 and assuming spherical spreading and negligible absorption, here is what we obtain:

$$EL = 234 - 40 \log R$$

$$RL_{CW} = 224 - 33 + 10 \log(40R) - 40 \log R = 207 - 30 \log R$$

$$RL_{FM} = 224 - 33 + 10 \log(0.13R) - 40 \log R = 182 - 30 \log R$$

$$NL = 70 - 3 - 18 = 49 \text{ dB}$$

The echo, noise and reverberation levels are plotted against range in Figure 13.2. The performance results using the plots are for a *single ping* and are therefore not directly comparable with the results from the full sonar equations derived above. Note in particular how excluding the 3 dB gain from multiple pings halves the reverberation-limited range. The effects of modifying parameters (TS, S_b, SL, B, N) can be easily and quickly seen by drawing new lines parallel to the appropriate existing lines. For example, increasing EL by 5 dB – by changing SL and/or TS – results in equal noise- and reverberation- limited ranges of 30 km.

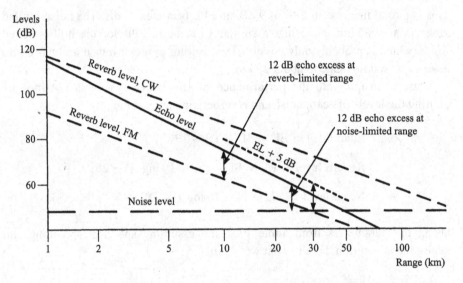

Figure 13.2   Echo, noise and reverberation levels plotted against range

13.8   Low Frequency Active: Beware!

As the frequency of operation of an active sonar is reduced, the noise-limited range will tend to increase because the reduction in absorption losses can outweigh the increase in the noise background (as demonstrated by the above examples). There are, however, two problems directly related to reducing the frequency and these can adversely affect performance:

- It is only possible to maintain a high SL over a limited bandwidth; it would be unwise to expect the usable bandwidth to be more than, say, one-third of an octave. The available bandwidth is therefore proportional to frequency. At frequencies of at least 2000 Hz, a bandwidth of 670 Hz (one-third of an octave) is realistic. This will meet the requirements for both detection and classification. At 300 Hz, however, the bandwidth will only be 100 Hz, sufficient for detection but not for classification.

- The frequency shifts due to target doppler (target motion relative to the sonar platform) are directly proportional to frequency. Target doppler is therefore inversely proportional to frequency. This is an unfortunate and important result for active sonars because it means that as the frequency is reduced to improve noise-limited detection range, there is an increase in the magnitude of target doppler needed to achieve noise-limited performance using a CW pulse. (Remember that the performance of a long CW pulse against a background of even low reverberation is always negligible.) Suppose a target echo needs a frequency shift of 2 Hz to avoid the main lobe of the reverberation spectrum (making any detection noise limited), this will be achieved by a target doppler of about 1 knot at a frequency of 3000 Hz but not until about 10 knots at 300 Hz, a very unlikely target doppler indeed!

13.9 Torpedo Detection

Modern torpedoes are frequently very quiet and, as we saw in the passive sonar examples, difficult to detect and classify in time to take evasive action or launch countermeasures. Active sonar can, with advantage, be used against such targets.

Reducing target strength produces a rapid fall in performance in even moderately high reverberation areas. If the background is predominantly noise, the fall will be much less. Fortunately, torpedoes are high doppler targets for most of the time and a well-chosen CW transmission will ensure that detection is against a noise background. Furthermore, the target doppler (without which a detection will not be made) gives an immediate classification, and because it is an active system, the location of the torpedo is known immediately.

Almost any active sonar able to detect submarines can be modified to detect torpedoes. As an example we will take the representative hull-mounted sonar and modify some of its parameters to get Table 13.3; the modified parameters are in italics.

Table 13.3 Modified parameters of hull-mounted surface ship sonar

Parameter	Value	Units
Frequency	*5.5*	kHz
Pulse length, T	*100*	*ms*
Source level, SL	224	dB
FM bandwidth	750	Hz
CW bandwidth	*20*	*Hz*
Horizontal beamwidth, θ_h	8	degrees
Reverberation index, RI		
FM	20	dB
CW	−6	dB
DI (receive)	25	dB
Target strength, TS	*−15*	*dB*
$5 \log d$	12	dB
$5 \log n$	3	dB
Background noise	55	dB
S_b	−33	dB

Modified parameters are in italics

Pulse length

The pulse length has been reduced to 100 ms. A 1 s pulse would have a dead range of 750 m, which is obviously too long for this application, and the range accuracy would also suffer. The 100 ms CW pulse will have an acceptable dead range of 75 m and the range resolution of the pulse will be somewhat better than this, say 20 m.

CW bandwidth

The CW bandwidth is therefore about 20 Hz $(2/T)$, giving a doppler cell width of about 3 knots. Therefore quite small dopplers (for a torpedo) will ensure noise-limited detections. For half-beam processing, about 5 knots will avoid the pulse main lobe; but for full-beam processing, target dopplers up to at least platform speed will be necessary to avoid reverberation in the beam sidelobes. Nevertheless, even this is unlikely to be a problem as torpedo dopplers will usually be greater than platform speed.

Target strength

The target strength is the suggested value for a random aspect torpedo, -15 dB.

Performance

The noise-limited performance of the sonar is given by the following equation:

$$2PL = SL + TS - N + DI + 10\log T - 5\log d + 5\log n$$

Substituting the values from Table 13.3, we obtain

$$2PL = 224 - 15 - 55 + 25 - 10 - 12 + 3 = 160 \text{ dB}$$

The allowable $PL = 80$ dB, and assuming spherical spreading plus absorption ($\alpha = 0.5$), the noise-limited detection range is about 7 km. This range is achieved for both the FM and CW pulses because they have the same duration.

The reverberation-limited performance of the sonar is given by the following equation:

$$10\log R = 10\log(B/\theta_h) - S_b - 41 + TS - 5\log d + 5\log n$$

Substituting the values from Table 13.3:

- For the FM pulse $10 \log R = 20 + 33 - 41 - 15 - 12 + 3 = -12$ dB and the reverberation-limited range is only 60 m

- For the CW pulse $10 \log R = -6 + 33 - 41 - 15 - 12 + 3 = -38$ dB and the reverberation limited range is only 0.2 m

Neither of these results should be surprising. For the FM transmissions the short range when compared with the submarine target is entirely due to the much lower target strength of the torpedo (25 dB less than the submarine value), and for the CW transmissions the negligible reverberation-limited ranges for either target are entirely due to the much lower reverberation indices of the CW pulses.

The optimum transmission for torpedo detection and classification, then, is a fairly long CW pulse (as long as possible consistent with dead range and resolution requirements). *But there must be some target doppler.* Because of its wide bandwidth, an FM pulse cannot use target doppler to avoid reverberation, so unless the reverberation background is very low, an FM pulse will not detect a torpedo at any target doppler.

Conclusion

The science of Underwater Sound had its beginnings in the 19th century – if we
ignore the oft quoted use of listening tubes by Leonardo da Vinci in the 15th.
However, the use of Underwater Sound to locate targets (Sonar) is comparatively
recent and the lineage of present day systems can be traced back to the Second
World War.

By the end of WW2, active sonars operating at around 20 kHz and using
transducer arrays of only a few elements, were able to detect submarines at ranges
of 1 or 2 km at best. The elements of the system were little different from today.

After WW2, improvements to submarines and torpedoes forced the sonar
designer to use lower frequencies to increase detection ranges and, in the fifties,
sonars like the RN Type 177 operating at between 6 and 12 kHz were entering
service in NATO and Warsaw Pact Navies. These sonars, when used in controlled
trials with alerted operators, were shown to be capable of ranges of up to about
20 km, but, operationally, this performance was seldom achieved. This *Operational Degradation*, which could result in detection ranges of only about 30% of
theoretical values, has been largely overcome by the use of computers first
introduced during the seventies to help the naval operator perform his demanding
surveillance and classification tasks.

During the final decades of the 20th century, ever quieter submarines and longer
range weapons have driven sonar designers to further reduce operating frequencies
– for both active and passive systems – resulting in the present generation of long
towed arrays and low frequency towed projectors.

What of the future? More decibels could be gained by, for example, increasing
SL and using long multiple arrays. Such steps would be very demanding in cost,
power and difficulties of deployment: current active sonars already use about
50 kW of electrical power, and more complex towed arrays for passive sonars
would present severe handling problems.

Sonar is a mature discipline that would appear to have reached fundamental –
or at least practical – limits: Active sonars will rarely detect submarines beyond

about 40 to 50 km range. Effort should not be squandered in near impossible attempts to extend this limit. Rather, the immediate future tasks should be:

(1) To ensure that this limit is more often reached

(2) To reduce false alarms in all environments

Neither task is easy. The first would be helped by exploiting favourable propagation modes – for example, VDS and Bottom Bounce (not only in an attempt to increase maximum range but also to fill in any intermediate shadow zones). The second would be helped by a unified approach to detection and classification; the careful choice of pulses and processing to best achieve this unified approach; and thorough integration of data on contacts from all available sources. Submarine *detection* needs to be optimised against both noise and reverberation, *but without compromising classification* (which needs the best possible resolutions in range, bearing and doppler).

It is the author's firm belief that these requirements are best met by a combination of an FM pulse (for best range resolution) and a CW pulse (for best doppler resolution), and half beam processing (for best bearing resolution).

Although much effort in recent years has gone into the design of alternative pulses in attempts to meet the above requirements with one pulse, it has not completely succeeded because; inevitably, any alternative pulse will compromise the best performance obtained from combined FM and CW pulses. Furthermore, careful study into the performances of alternative pulses against all possible dopplers will frequently reveal significant ambiguities.

Despite the clear advantages of half beam processing, there are two arguments frequently used against it:

(1) The half beam DI is 3 dB less than the full beam DI. Note, however, that the bulk steer and add process after detection restores most or all of this reduction (compare with passive broadband energy and cross correlator detection, Sections 8.12 and 8.13). In any event the gains in the unified detection and classification task from having additional information on the bearing spreads of targets and reverberation will more than compensate for this loss.

(2) The success of the phase comparison process relies upon a good signal-to-noise ratio. But this is inevitably the case for an active sonar system where marginal (low SNR) detections are not a consideration: typically there will be no discernible echo on one ping and a strong echo on the next, due mainly to

fluctuations in target strength caused by perhaps quite small changes in target aspect from ping to ping. (Contrast with passive systems where the signal energy is – in the short term at least – constantly available and a track will gradually, at low SNR, become apparent against the noise background.)

Finally, dear reader, I hope that you have enjoyed reading my book as much as I have enjoyed writing it. I remain dedicated to the advancement of sonar – which, in these cost conscious days, may simply be to arrest any backward move! – and would welcome dialogue with interested readers, particularly those who point out the inevitable errors for correction in future impressions.

Solutions to Problems

Problem 1.1

$I = p^2/\rho c = (10^{-4})^2/(1.5 \times 10^6) = 0.67 \times 10^{-14}\,\text{W/m}^2$

$I_r = 0.67 \times 10^{-18}\,\text{W/m}^2$

$I/I_r = 10\log 10^4 = 40\,\text{dB}$

Problem 1.2

$SL = 10\log P + 170.8 + DI_t$

$\quad = 10\log 40\,000 + 171 + 15$

$\quad = 232\,\text{dB}$

Problem 1.3

From Figure 1.2 the safe power density is 5 kW/m². This may be doubled because of the operating frequency. The radiating surface is $2 \times 1 \times (\pi/4) = 1.6\,\text{m}^2$. Therefore the total safe radiated power is $2 \times 5 \times 1.6 = 16$ kW. The factor $\pi/4$ is needed to calculate the actual radiating surface resulting from the use of circular elements.

Problem 1.4

$BL = SpL + 10\log\Delta f$

$\Delta f = 2000\,\text{Hz}$ and $BL = 80\,\text{dB}$

Therefore $SpL = 80 - 10\log 2000 = 47\,\text{dB}$

Problem 2.1

The 3 dB beamwidth of a line array is given by

$$2\theta_3 = \frac{76}{Lf}\left(1 + \frac{\theta_s^2}{4000}\right)$$

where L is length in metres, f is frequency in kHz and θ_s is the steer angle.

(i) $2\theta_3 = 76/(10 \times 4) = 1.9°$

(ii) $2\theta_3 = [76/(10 \times 4)](1 + 0.9) = 3.6°$

Note that a steer of 60° has almost doubled the beamwidth.

Problem 2.2

The sidelobe levels are given by $20\log[2/\pi(2m + 1)]$. For $m = 4$ the sidelobe levels are 23 dB below the main lobe. Their angular positions are given by $(\pi L/\lambda)\sin\theta = \pm 9\pi/2$, and since $L = 6\lambda$ we have $\theta = \pm 49°$. Note that the sidelobe levels do not depend on the dimensions of the array. The higher-order sidelobes only exist, however, if the array is large enough (in terms of wavelengths) to generate them

Problem 2.3

At 10 kHz, $\lambda/2 = 75$ mm. The array therefore has $2000/75 = 27$ rows of $5000/75 = 67$ elements. At 8 kHz we have

$$DI = 3 + 10\log(27 \times 67) - 20\log(10/8) = 33.7 \text{ dB}$$

Alternatively

$$DI = 3 + 10\log(4Lh/\lambda^2) = 10\log(4 \times 5 \times 2/0.1875^2) = 3 + 30.6 = 33.6 \text{ dB}$$

The trivial difference is due to the element numbers being the nearest integer values.

Problem 2.4

The DI of a baffled cylinder is given by

$$DI = 10\log 5hdf_0^2 = 10\log(2 \times 3 \times 5^2) = 22 \text{ dB}$$

Halving the height of the array reduces the transmit DI by 3 dB. The number of elements in the array is halved and therefore the transmitted power is also reduced by 3 dB. The source level is therefore reduced by a total of 6 dB.

Problem 3.1

(i) The equation for spherical spreading plus absorption is

$$PL = 20 \log r + (ar \times 10^{-3})$$

At 5 kHz $80 = 20 \log r + (0.3r \times 10^{-3})$ therefore $r = 7$ km

At 20 kHz $80 = 20 \log r + (3r \times 10^{-3})$ therefore $r = 3.3$ km

(ii) The equation for cylindrical spreading plus absorption is

$$PL = 10 \log r + (ar \times 10^{-3})$$

At 5 kHz $80 = 10 \log r + (0.3r \times 10^{-3})$ therefore $r = 100$ km

At 20 kHz $80 = 10 \log r + (3r \times 10^{-3})$ therefore $r = 13$ km

A value of PL = 80 dB is realistic for many sonar systems, and so are the ranges obtained by assuming spherical spreading plus absorption. The ranges obtained by assuming cylindrical spreading plus absorption, however, are usually far too large in practice. Such ranges would only be achieved using the deep sound channel (DSC) mode where the sound is confined to cylindrical spreading by refraction alone. For all other modes there will be losses at the boundaries, and these losses will significantly reduce the range.

Problem 4.1

The target strength of a sphere is given by $TS = 10 \log(a^2/4)$. Therefore $TS = 10 \log(0.5^2/4) = -12$ dB. The minimum dimension of the sphere is 1 m, therefore the maximum wavelength for a reliable value of $TS = 0.2$ m. The frequency must therefore be at least $c/\lambda = 1500/0.2 = 7500$ Hz.

Problem 4.2

The target strength of a cylinder is given by $TS = 10 \log(aL^2/2\lambda)$:

At 10 kHz $TS = 10 \log[(0.5 \times 2^2)/(2 \times 0.15)] = -2$ dB

At 100 kHz $TS = 10 \log[(0.5 \times 2^2)/(2 \times 0.015)] = 8$ dB

End on, the target strength is that of a sphere, $TS = 10 \log(a^2/4)$. The TS does not change with frequency (provided the diameter is at least 3λ) and therefore, for both frequencies, $TS = 10 \log(0.5^2/4) = -12$ dB.

Problem 4.3

The target strength of any plate at an angle θ to normal is given by

$$TS = 10 \log(A/\lambda)^2 + 20 \log(x^{-1} \sin x) + 20 \log(\cos \theta)$$

The largest dimension is $2\,\text{m}$ (replacing L), so we use $x = (2/\lambda)\sin \theta = (2/0.375)\sin 5° = 0.46$ and, from the dashed plot of Figure 4.1, we have $20 \log(x^{-1} \sin x) = -16$ dB. Also $20 \log(\cos 5°) = 0$ dB, therefore

$$TS = 10 \log(5/0.375)^2 - 16 + 0 = 22 - 16 + 0 = 6 \text{ dB}$$

Problem 5.1

At 100 kHz there will be two noise sources: thermal noise and ambient sea noise. The thermal noise is given by $-15 + 20 \log f$ (f in kHz), therefore $N_{\text{thermal}} = 25$ dB. Ambient noise, from Figure 5.1, is also 25 dB. So the total noise is $25 + 3 = 28$ dB. The output voltage from the hydrophone is given by

$$20 \log v = S_h + 20 \log p + 120 \qquad \text{(adding 120 dB puts } v \text{ in } \mu\text{V)}$$

$$= -170 + 28 + 120 = -22 \text{ dB}$$

To find the output in a 1000 Hz band, add $10 \log 1000 = 30$ dB, so $20 \log v = 8$ dB and $v = 2.5\ \mu\text{V}$.

Problem 5.2

The isotropic spectrum level of the noise at the array/water interface is $55 + 3 = 58$ dB because self-noise $=$ ambient noise $= 55$ dB. One third-octave is 2000 Hz, therefore the isotropic band level is $58 + 10 \log 2000 = 91$ dB. The noise measured at the output of the receiver beam is $91 - 20 = 71$ dB. Because self-noise measurements are traditionally made at the beam outputs of a sonar receiver, they are sometimes misleadingly quoted without adding the DI of the beam. So beware of any suspiciously low values for self-noise.

Problem 6.1

Start with

$$TS_R = S_s + 10 \log A \qquad A = \frac{cT}{2} R\theta_h$$

where R is in metres and θ_h is in radians. At 4000 m the reverberating area is

$$A = (1500 \times 10^{-1}/2) \times 4000 \times 0.17 = 51\,000 \text{ m}^2$$

$$\text{TS}_R = -40 + 47 = 7 \text{ dB}$$

Doubling the range doubles the reverberating area, therefore

$$\text{TS}_R = 7 + 3 = 10 \text{ dB}$$

The TS of a vessel is constant with range. Suppose the echo needs to be 10 dB greater than the background reverberation to be detected, then for a TS of 20 dB, the system would be reverberation limited at 8000 m. At greater ranges, because the TS_R increases, reverberation would mask the vessel and it would not be detected.

Problem 7.1
Using $P_d = 0.5$, we can read off P_{fa}:

$$\text{For } 5 \log d = 6 \text{ dB}, \ P_{fa} = 2 \times 10^{-5}$$

$$\text{For } 5 \log d = 5 \text{ dB}, \ P_{fa} = 10^{-3}$$

Therefore P_{fa} is increased 50 times. Note how a small reduction in the detection threshold (of which $5 \log d$ is one of the terms) results in a large increase in false alarms. This will have important implications for an automatic detection system.

Problem 8.1
The only parameter to change is DI:

At 3000 Hz	$\text{DI} = 3 + 10 \log 50 + 20 \log 1.5 = 23 \text{ dB}$
At 400 Hz	$\text{DI} = 3 + 17 - 20 \log 5 = 6 \text{ dB}$
At 200 Hz	$\text{DI} = 3 + 17 - 20 \log 10 = 3 \text{ dB}$
At 80 Hz	$\text{DI} = 3 + 17 - 20 \log 25 = 3 \text{ dB}$
At 40 Hz	$\text{DI} = 3 + 23 - 20 \log 50 = 3 \text{ dB}$

Because the flank arrays are baffled by the hull, DI cannot be less than 3 dB. The PL and R values are then as follows.

Broadband

PL, $\alpha = 0.2$ (dB)	128	98	93
R (km)	>100	35	25

Narrowband

PL (dB)	89	86	59	56	64	61
R (km)	30	20	0.9	0.6	1.6	1.1

Broadband detection performance is still very good, even with the much smaller flank arrays. Narrowband performance is still good for the noisy torpedo, poor but unchanged for the submarine, and still further reduced for the quiet torpedo. Had the size of the array been doubled, which would hardly be practical, performance would still be unsatisfactory against the quiet targets, thus demonstrating the need for long towed arrays.

Problem 8.2

The time difference between the two paths is given by

$$\delta T = \frac{4Hp}{c(R^2 + 4H^2)^{1/2}}$$

This solves to give $R = 15\,490$ m. Note that because neither path includes a component due to a surface reflection from the target, t does not appear in the equation. To find t one of the paths must include a surface reflection from the target.

Problem 8.3

$DI = 10 \log n = 21$ dB, therefore the array must have 126 elements. They will be spaced $\lambda/2$ at 3000 Hz. $\lambda/2 = c/2f = 1500/6000 = 0.25$ m, and the array length is $126 \times 0.25 = 31.5$ m. At 3000 Hz, $PL = 20 \log 10\,000 + (0.2 \times 10\,000 \times 10^{-3})$ $= 82$ dB. The sonar equation to use is

$$PL = SL - N + DI - 5 \log d + 5 \log BT_e + 5 \log n$$

Therefore

$$SL = 82 + 60 - 21 + 6 - 22 - 9 = 96 \text{ dB}$$

This is the spectrum level of the radiated noise of a fairly quiet submarine, implying that a quite short towed array may be adequate to detect submarines. The problem, however, is classification and the need to discriminate against other, probably noisier, targets such as surface ships, for which a capability at much lower frequencies is essential.

Problem 9.1

The frequency shift due to ship (platform) motion is $\Delta f = 2S \cos \theta \times (f/c)$. Be careful with the units: S must be in m/s if c is in m/s.

$$\Delta f = 2 \times 4 \times 0.707 \times (5000/1500) = 18.8 \text{ Hz}$$

The total shift is 25 Hz, therefore the shift due to the target is 6.2 Hz:

$$6.2 = 2 \times (S \cos \varphi) \times (5000/1500)$$

Therefore $S \cos \varphi = 0.93$ m/s or 1.8 knots. This is the relative velocity of the target (its doppler); we cannot separate the target speed from its bearing without more information.

Problem 9.2

There will be a 3 dB processing loss when the replica only matches the echo for 0.7 of its bandwidth. This will result from a 120 Hz frequency shift:

$$\Delta f = 2 \times (S \cos \varphi) \times (f/c)$$

$$120 = 2 \times (S \cos \varphi) \times (4000/1500)$$

$$S \cos \varphi = 22.5 \text{ m/s or 45 knots}$$

There will be a 1 dB processing loss when the replica matches the echo for 0.9 of its bandwidth. This will result from a 40 Hz frequency shift:

$$40 = 2 \times (S \cos \varphi) \times (4000/1500)$$

$$S \cos \varphi = 7.5 \text{ m/s or 15 knots}$$

Even 15 knots is a lot of doppler for a submarine (remember that doppler is relative velocity, not target speed). Therefore there is seldom any point in using an extended reference.

Problem 9.3

For a Hamming shaped pulse, $\Delta f_{40} = 3.5/T$. Therefore the doppler shift must be ± 1.9 Hz to achieve $R_j = -40$ dB.

At 3000 Hz

$$1.9 = 2 \times \text{target doppler} \times 3000/1500$$

$$\text{target doppler} = 0.475 \text{ m/s or about 1 knot}$$

At 1000 Hz

$$1.9 = 2 \times \text{target doppler} \times 1000/1500$$

$$\text{target doppler} = 1.425 \text{ m/s or about 3 knots}$$

At 300 Hz

$$1.9 = 2 \times \text{target doppler} \times 300/1500$$

$$\text{target doppler} = 4.75 \text{ m/s or about 10 knots}$$

Doppler shift is proportional to frequency. Target doppler is therefore inversely proportional to frequency, an unfortunate and important result for active sonars because it means that as the frequency is reduced to improve detection range, it increases the magnitude of target doppler needed to achieve noise-limited performance when using a CW pulse.

Problem 10.1

Make a guess and then iterate. Suppose $f = 20$ kHz. Propagation loss (dB) will be given by

$$PL = 20 \log r + \alpha r \times 10^{-3}$$

$$= 20 \log 12\,000 + 12\alpha = 82 + 12\alpha \tag{1i}$$

For noise-limited detection,

$$2PL = SL + TS - N + DI + 10 \log T - 5 \log d + 5 \log n$$

$$= 210 + 40 - N + 26 - 20 - 10 + 3 = 249 - N \quad \text{(dB)}$$

At 20 kHz, $N = 40$ dB therefore PL = 105 dB. Equating to (i), we have $105 = 82 + 12\alpha$, so $\alpha = 2$. Clearly, given $T = 10$ ms then 20 kHz is too high ($\alpha = 3.8$ at 20 kHz). We would need to reduce the frequency to about 15 kHz. Such a comparatively low frequency implies a large, expensive array and a better solution would be to use a long broadband pulse.

Try a 1 s, 20 kHz, 100 Hz bandwidth, FM pulse. The resolution would be the same but now

$$2PL = 210 + 40 - 40 + 26 + 0 - 10 + 3 = 229 \text{ dB}$$

Again, equating to (i), $115 = 82 + 12\alpha$ therefore $\alpha = 2.8$ and the frequency is now about 18 kHz. Note the significance of absorption at these higher frequencies: it is not practical to achieve, say, 10 km range at frequencies in excess of about 15 kHz, even given a very strong target like an ocean floor at normal incidence. At 30 kHz we have

$$PL = 115 = 20 \log r + 7r \times 10^{-3} \quad (dB)$$

therefore $r = 5700$ m. The TS of the bottom is given by $TS_B = S_B + 10 \log A$. At normal incidence S_B will be at least -20 dB, and at a depth of 12 000 m and for a beam of solid angle $5°$ then $10 \log A = 60$ dB. Therefore $TS_B = -20 + 60 = 40$ dB. But this target strength does assume the bottom to be flat (within 7.5 m for the effective pulse duration of 10 ms) throughout the area A.

Problem 11.1
Noise limited

The noise-limited active sonar equation is

$$2PL = SL + TS - N + DI + 10 \log T - 5 \log d + 5 \log n$$

For the FM pulse

$$2PL = 192 - 25 - 30 + 43 - 20 - 10 + 3 = 153 \text{ dB}$$

therefore PL = 77 dB:

$$PL = 77 = 20 \log r + \alpha r \times 10^{-3} \quad (\alpha = 50 \text{ dB/km})$$

giving $r = 470$ m.

For the CW pulse

$$2PL = 192 - 25 - 30 + 43 - 39 - 10 + 3 = 134 \text{ dB}$$

therefore PL = 67 dB:

$$PL = 67 = 20 \log r + \alpha r \times 10^{-3} \quad (\alpha = 50 \text{ dB/km})$$

giving $r = 330$ m.

Reverberation limited

The reverberation-limited active sonar equation is

$$10 \log R = 10 \log(B/\theta_h) - S_b - 11 + TS - 5 \log d + 5 \log n$$

where R is the range in metres.

For the FM pulse

$$10 \log R = 49 + 30 - 11 - 25 - 10 + 3 = 36 \text{ dB}$$

Therefore $R = 4000$ m. The reverberation-limited range requirement (to be greater than the noise-limited range) is easily met with this pulse.

For the CW pulse

$$10 \log R = 39 + 30 - 11 - 25 - 10 + 3 = 26 \text{ dB}$$

Therefore $R = 400$ m. The reverberation-limited range requirement (to be greater than the noise-limited range) is met with this pulse.

Increasing the TS by 5 dB meets the original detection range requirement using either pulse. The higher resolution of the FM pulse, however, means that it is still the best pulse for the total task of detection and classification.

Problem 11.2
Use the approximate formula for absorption attenuation, $\alpha = 0.05 f^{1.4}$, and assume spherical spreading.

Frequency (kHz)	60	80	100
α (dB/km)	15	23	32
PL (dB)	70	76	84
SL (dB)	205	205	205

The noise-limited performance is given by

$$2PL = SL + TS - N + DI + 10 \log T - 5 \log d + 5 \log n$$

At 100 kHz we obtain

$$2PL = 168 = SL - 30 - 35 + 45 - 10 - 10 + 3$$

and SL = 205 dB. To flatten the response at the receiver, the attenuation at 60 kHz must be $2 \times (84 - 70) = 28$ dB.

This equalization is only correct at 800 m. In practice it would probably be better to equalize at, say, half maximum range (400 m) and accept the smaller changes in response at other ranges. SL will change with frequency and the equalization should also take this into account.

Problem 12.1

The intercept sonar equation is

$$PL = SL + DI - N - 5 \log d - 10 \log B_r$$

For SL = 180 dB we have

$$PL = 180 + 10 - 40 - 10 - 10 \log 10\,000 = 100 \text{ dB}$$

At 15 kHz, $\alpha = 1.6$ and

$$PL = 20 \log r + 1.6r \times 10^{-3} = 100 \text{ dB}$$
$$r = 12\,000 \text{ m}$$

For SL = 220 dB we have

$$PL = 220 + 10 - 40 - 10 - 10 \log 10\,000 = 140 \text{ dB}$$
$$PL = 20 \log r + 1.6r \times 10^{-3} = 140 \text{ dB}$$
$$r = 31\,000 \text{ m}$$

Note that the pulse type and duration is irrelevant to intercept (provided its bandwidth is less than B_r.

Problem 12.2

The active sonar noise-limited equation is

$$2PL = SL + TS - N + DI + 10 \log T - 5 \log d + 5 \log n$$

For SL = 180 dB we have

$$2PL = 180 + 15 - 40 + 20 + 0 - 10 + 3 = 168 \text{ dB}$$
$$PL = 20 \log r + 1.6r \times 10^{-3} = 84 \text{ dB}$$
$$r = 5000 \text{ m}$$

For SL = 220 dB we have

$$2PL = 220 + 15 - 40 + 20 + 0 - 10 + 3 = 208 \text{ dB}$$

$$PL = 20 \log r + 1.6r \times 10^{-3} = 104 \text{ dB}$$

$$r = 13\,000 \text{ m}$$

As always, this (noise-limited) performance will only be achieved if the target submarine shows sufficient doppler. Note that the 40 dB (100 times) increase in transmitted power only increases the range by a factor of 2.6.

Index